Symmetry:

De Rerum Structura

Carlo Faustini

DEDICATION

To my family for their support and encouragements.

To Angelo Cacciavillani who taught me the art and science of problem solving.

CONTENTS

ACKNOWLEDGMENTS

This book embodies more than half a century of efforts. It started as a part time, pencil and paper project but it kept expanding at a rate that regularly required current state of the art personal computers with accompanying expenditures of time and money. I thank my departed wife Mirella and my whole family for supporting and encouraging me to persevere. I also thank my son Lou for his invaluable help going through computer upgrades, finding lost files and getting things moving again whenever I got stuck.

I thank Create Space for providing me the tools to get a head start in publishing the book and for their excellent job editing my manuscript. I also thank my daughter in law Andrea and the whole publishing team for their patience dealing with my severely impaired hearing.

CHAPTER 1

ASSESSING THE PROBLEM

Introduction

To escape from a seemingly unending stretch of trivial work, I looked for a personal project that would involve a modicum of original research.

In the last years of my high school, I became very familiar with the development, characteristics, and use of the quadratic formula. I learned that a formula also existed for the cubic. During my undergraduate years, I came across problems that required finding the roots of high-degree equations that no formulas existed for. So I set finding a formula for the quartic as my project goal.

To keep original research as the project's primary purpose, I deliberately avoided previous work on the subject. To broaden the project scope, I expanded the goal to include searching for a generic methodology that could be applicable to equations of any degree. I spent more than a decade and considerable effort searching for the essential characteristics that define the problem in order to ensure that I sought solutions for the real problem and not a *preferential solution*—for instance, the operation of extracting the fifth root of a real number produces five roots, one of which is real, while all the others are complex numbers. The real root is the preferential solution because, in most cases, it is the one being sought and the easiest one to express using familiar symbols. The others are most commonly disregarded as not being of interest.

I also spent time and effort choosing the degree for the first equation to be solved. After I rummaged through my bag of tricks, I chose the cubic because I had some vague ideas regarding how to go about it, and if I was successful, the results could be compared with those of the existing formula.

The initial investigation produced three different ways of looking at the problem:

1. Finding the roots of a generic cubic
2. Using a set of relationships known as the *elementary symmetrical functions*, or ESFs $\{\sigma_1, \sigma_2, \sigma_3\}$, that directly bind the roots of the equation to its coefficients
3. Exploiting the one characteristic that binds the roots directly to themselves

Method 1: Finding the Roots of a Generic Cubic

Given the cubic

$$y = x^3 + ax^2 + bx + c$$

find its *three* roots $\{x_1, x_2, x_3\}$, that is the values of x for which y equals to zero.

This method limits the scope of the problem to what is essentially a preferential solution. Operations performed in this method are aimed at finding a formula, and they involve only the coefficients of the cubic.

Method 2: Using ESFs $\{\sigma_1, \sigma_2, \sigma_3\}$ That Directly Bind the Roots of a Cubic to Its Coefficients

The ESFs of a generic cubic and their relationships to the cubic coefficients are

$$\sigma_1 = x_1 + x_2 + x_3 = -a$$

$$\sigma_2 = x_1 x_2 + x_2 x_3 + x_3 x_1 = b$$

$$\sigma_3 = x_1 x_2 x_3 = -c$$

These relationships can be used in two ways, depending on whether the roots have already been found or are yet to be determined. In the first case, they have been used to determine which three values out of the possible nine generated by the old formula are the roots of the cubic of interest (COI). In the second case, they could be considered as a sixth-degree system of equations. My recollection about the solutions was that each variable could have six possible values. How they would relate to one another and to the three roots of the COI could only be speculated. This second case might also require solving a sixth-degree equation, which is two degrees above the original goal and no formula exists for. Pursuing it would have meant abandoning the initial goal.

Operations to be performed with this method would require using both the cubic roots and its coefficients.

Method 3: Exploiting the One Characteristic That Binds the Roots Directly to Themselves

That one characteristic can be described by a single word—*symmetry*. This appeared to be the most direct way to solve the problem and seemed worth pursuing.

Operations performed with this method involve only the cubic's roots and generate symmetric expressions thereof. They would not involve the cubic's coefficients. The cubic's ESFs would then be used to translate the symmetric expressions into functions of the cubic coefficients by means of specifically designed procedures.

The task then became to define *symmetry* and turn it into a tool for accomplishing the goal. After several unsuccessful searches for an academic definition, I chose two from *Merriam-Webster's Collegiate Dictionary*. The first was as follows:

> **symmetry:** the property of being symmetrical; *especially*: correspondence in size, shape, and relative position of parts on opposite sides of a dividing line or median plane or about a center or axis.

The second was as follows:

> **symmetrical or symmetric:** being such that the terms or variables may be interchanged without altering the value, character, or truth.

The first definition appears better suited for geometric figures and the second for algebraic expressions. Together, they were instrumental in assembling the roots as parts of partially symmetric structures as the first steps in creating a formula.

The first definition of symmetry requires not only a plurality of parts but also something for the parts to be symmetrical about and rules to specify their correspondence. In addition, symmetry is defined as a *property*, a characteristic that needs to be associated with a noun, something to completely define its meaning. *Structure* was chosen as that something. The *Merriam-Webster's Collegiate Dictionary* defines it as

> **structure:** something made up of interdependent parts arranged in a definite pattern of organization.

Symmetry, then, would act as the property characterizing the definite pattern of organization of the parts within the structure.

Cubic Roots of Unity

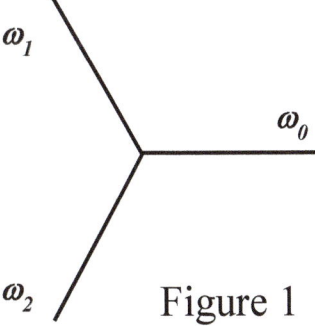

Figure 1

The cubic roots of unity play a paramount role in creating the definite patterns of organization needed for assembling the roots of a generic cubic as parts of a symmetrical structure. Figure 1 depicts the most important properties that are relevant to this project.

Geometrically, the roots look like a triad of polar vectors with unit magnitudes and angles respectively of 0°, 120°, and 240°. For reference purposes, these vectors are designated individually as ω_0, ω_1, and ω_2 and collectively as *principal roots*. They are characterized by angles that are less than 360°. These vectors are not the only possible ones because vectors that are congruent to them modulo 360° also produce triads that look identical to that in figure 1.

The vectors can be expressed as complex numbers or as polar coordinates, or can be given a generic exponential form as follows:

$$\omega_n = \varepsilon^{j2\pi n}; \qquad \omega_{n+1} = \varepsilon^{j2\pi\left(n+\frac{1}{3}\right)}; \qquad \omega_{n+2} = \varepsilon^{j2\pi\left(n+\frac{2}{3}\right)}$$

The letter j stands for the imaginary unit. The n parameter in the exponent is limited to integer values, and it determines the behavioral characteristics of the $\{\omega_0, \omega_1, \omega_2\}$ set. If n equals 0, the result is the set of *principal roots*. The result of choosing n equal to any other integer is to rotate the associated principal roots n times by 360°. Counterclockwise (CCW) rotations are defined as positive and correspond to positive exponents; clockwise (CW) rotations are defined as negative and correspond to negative exponents. Choosing n as a sequence of integers continuously incrementing (or decrementing) by 1 has the effect of continuously stepwise rotating the associated principal root in 360° increments. There is no apparent change to its angular location. Interpreting the n parameter as a sequence of time instants endows the roots with a (rather hidden) dynamic behavior.

The results of choosing non-integer values for n are and should be deemed as not pertinent because the outcomes are not cubic roots of unity. A root dynamic behavior is then to be considered as a stepwise, rather than a continuous, motion.

The effect is analogous to using a stroboscopic light to look at a rotating wheel with marks on it. If the light flashes are not synchronized with the wheel's angular velocity, the results are meaningless and are analogous to choosing non-integer values for n. Assume that the wheel has three identical and equally spaced marks and that the flashes occur once per revolution and are locked to one mark. The result is that only that mark would be seen and the other two would remain hidden. If the flashes occur three times per revolution, are equally spaced in time, and are locked to any one mark, all three marks, although seen sequentially, would appear as a single mark. This situation is analogous to replacing the expression for ω_0, ω_1, and ω_2 with a common one that covers all three.

$$\omega = \varepsilon^{j2\pi(n/3)}$$

An interesting and productive way of expressing the cubic roots of unity follows:

$$\rho_0 = \varepsilon^{j\frac{2\pi n}{3}}; \qquad \rho_1 = \varepsilon^{\frac{j2\pi}{3}(n+1)}; \qquad \rho_2 = \varepsilon^{\frac{j2\pi}{3}(n+2)}$$

Assuming n as a sequence of integers continuously incrementing (or decrementing) by one has the effect of continuously rotating the associated root stepwise by 120° increments—CCW for increments and CW for decrements. For increments, each root circularly moves into the position previously occupied by the root at its right—that is, ρ_0 moves into the ρ_1 position, ρ_1 moves into the ρ_2 position, and ρ_2 moves into the ρ_0 position. The following table shows the angular position of the $\{\rho_0, \rho_1, \rho_2\}$ triplet as a function of n.

n	ρ_0	ρ_1	ρ_2
0	$0°$	$120°$	$240°$
1	$120°$	$240°$	$0°$
2	$240°$	$0°$	$120°$

Values of n that are congruent to those in the table modulo 3 produce the same angular positions.

As a three-part symmetrical structure, the triplet $\{\rho_0, \rho_1, \rho_2\}$ geometrically looks the same as the triplet $\{\omega_0, \omega_1, \omega_2\}$, except that the parts of the former appear to be moving from one location to another, while the parts of the latter appear as being newly created in the same location.

The following properties of the principal roots of unity will be frequently used herein:

1. Their sum is equal to 0—$\omega_0 + \omega_1 + \omega_2 = 0$.
2. Their products can be performed by adding subscripts modulo 3—for example, $\omega_1 \times \omega_2 = \omega_0$; $\omega_1 \times \omega_1 = \omega_2$; $\omega_2 \times \omega_2 = \omega_1$.
3. The result of multiplying a polar vector by a principal root is to rotate it by an angle equal to that associated with the principal root. In this role, principal roots function as mathematical operators, and when used as such, they will be referred to herein as *rotators*. The imparted rotation can be treated as a single step or as a sequence thereof.
4. The paramount property of ω_0, ω_1, and ω_2 as a triad is that they form a *symmetrical structure* with respect to their center. The correspondence rule can be stated as "every root of the triad has a correspondent one rotated by $2\pi/3$ radians or *any integer multiple thereof* in either direction around their center." Properly modified, this rule also applies to larger structures assembled by using them as rotators.

For the sake of brevity, this type of correspondence may be shortened to "the parts have a $2\pi/3$ (or 120°) correspondence." The parts can be arithmetic expressions or elements, such as roots or structures.

Roots As Objects

A productive way to use roots is to treat them as objects in accordance with the principles of object-oriented technology as it is practiced in computer science. The technology was still evolving, and there appeared to be some variations of terminology and definitions among authors, especially those from different fields of science. I needed definitions that would retain their validity, regardless of possible changes in the state of the art. Therefore, the definitions used herein reflect my own interpretation of the technology at the time I was working on this part of the project (several decades ago). They may not be current nor agree completely with those of other users. The definitions given here are for the sole purpose of trying to clarify how they are used in this project by relating concepts to terminology.

The major attraction of object-oriented technology is that, once properly defined, an object can be reused and expanded without having to redefine it over and over again.

An object is defined as a collection of elements (usually data and variables) together with their properties and the rules by which they are to be operated upon. Objects are characterized by three processes:

1. *Encapsulation*: a process that combines the elements, properties, and rules of the object into one, closed, indivisible whole.
2. *Inheritance*: a process that allows increasing the number and types of elements, properties, and rules of an object (designated as the *ancestor*) by creating a new one (designated as the *descendant*) in which all the ancestor items (elements, properties, and rules) are retained unaltered, while new ones are added in with their own specifications.
3. *Polymorphism*: a process that allows manipulation of the elements inherited from the ancestor according to the inherited or to new superseding rules.

For ease of reference, objects that involve radicals with index N will be designated herein as *N-clusters*. Their relevant characteristics are described subsequently.

N-Clusters

N-clusters are defined as objects the elements of which are as follows:

1. A positive integer N that determines the number of roots in the object, the exponent to which a root has to be raised to obtain the power, and the radical index to be used when extracting roots from the power.
2. A *power element* determined by its amplitude A and angle α.
3. N root elements determined by a common amplitude R and angles $\rho_1, \rho_2,..., \rho_N$, which are circularly spaced by $2\pi/N$ radians or *any multiple thereof* in either direction. The amplitude of R is the Nth root of the A amplitude.

4. A cluster center of symmetry about which the roots are symmetrical. The roots and their center will be referred to as forming a *local symmetrical structure*.

Notes:
1. Some elements bear specific relationships to one another, and as such, they cannot be arbitrarily assigned. Also, since they can be computed in terms of other elements, not all of them need to be explicitly included in the object.
2. Elements can be constants, variables, arithmetic expressions, or local symmetrical structures.

The operations for manipulating an *N*-cluster element are as follows:

1. All four basic arithmetic operations—adding, subtracting, multiplying, and dividing.
2. Raising a root to the *N*th power—the result is always the same power element regardless of which root is chosen as the operand.
3. Extracting the *N*th roots from the power element—this operation generates all *N* roots at once as an inseparable group (actually as a local symmetrical structure). In normal practice, one root (which may not be the one that was chosen to compute the power) is selected as *the* root. The practice ignores and hides the paramount characteristic that roots are interchangeable parts of a symmetrical structure.
4. Geometrically, the roots appear as radii of a circle with contiguous ones spaced apart by $2\pi/N$ radians. Together, they form a symmetrical structure. If *N* is a prime number, the correspondence rules can be stated as, "For every root there is a correspondent one spaced by $2\pi/N$ radians (or an integer multiple thereof) apart." For the sake of brevity, such correspondence rules may be shortened to "the parts have a $2\pi/N$ (or 360°/*N*) correspondence." The parts can be root elements, structures, algebraic expression, or local symmetrical structures.

Raising a root to the *N*th power and extracting the *N*th roots from the power form an inverse pair of operations, with the latter producing a multiplicity of equally valid results. This makes it incompatible with the Group Theory rule, which requires that the result of every operation in the group be unique. Because all formulas for nonlinear equations require extracting roots, they cannot be regarded as members of the Group Theory (a.k.a. linear algebra) domain.

Note that N-clusters are defined solely on the basis of their geometric characteristics and independently of any algebraic coordinate system. In particular, no root element is assigned special or unique properties.

The following two characteristics of *N*-clusters are worth emphasizing:

1. *N*-clusters have many features and rules very similar to those of the cubic roots of unity. Considering the latter as a 3-cluster, its radical index 3 is generalized by replacing it with the integer *N* for both the radical index and in the rule that

defines correspondence among roots.

2. Notice that whereas any one multiple of $2\pi/N$ radians (with N prime) completely defines *the definite pattern of organization* for the symmetrical structure considered as a static configuration, a sequence of continuously incrementing integers is needed to describe its dynamic behavior. This statement will become more meaningful when describing the symmetrical structures associated with the roots of a multiplicity of cubic equations.

If N is not a prime number, the *correspondence rule* becomes more complicated. This situation has been considered for $N = 4$, while developing the formula for the quartic equation. It shows features, some of which are present in the cubic development and others that are typical for values of N that are not prime numbers.

The case for $N = 5$ comes up when developing the formula for the quintic. It shows features, some of which are present in the cubic and in the quartic development and some that are new.

These cases will be the subject of future books. Both expand the concepts of *the definite pattern of organization* and show why more than one may exist.

They also show the usefulness of object-oriented technology in developing structures that are similar to but with more features than those that will be described herein.

CHAPTER 2

ASSEMBLING SYMMETRIC STRUCTURES

Project Outline

The strategy for creating a cubic formula follows three steps:

1. Start with algebraic symmetrical structures that are functions of both cubic roots and rotators.
2. Follow up with a sequence of procedures that systematically transforms portions of these structures from functions of roots and rotators to symmetrical expressions of roots alone.
3. Use a specifically designed generic procedure to rewrite the symmetrical expressions of the roots in terms of the COI ESFs and, hence, in terms of the COI coefficients.

The terms of all symmetrical expressions are easily organized as geometric symmetrical structures with parts that can be either the roots or the coefficients of the COI or combinations thereof.

Symmetrical Structures of Three Variables: DyCon3

The following relationships are identities that are true for all values of the variables $\{x_1, x_2, x_0\}$.

$$(\omega_0 x_1 + \omega_1 x_2 + \omega_2 x_0) + (\omega_0 x_1 + \omega_1 x_0 + \omega_2 x_2) + (\omega_0 x_1 + \omega_0 x_2 + \omega_0 x_0) = 3x_1$$
$$(\omega_0 x_2 + \omega_1 x_0 + \omega_2 x_1) + (\omega_0 x_2 + \omega_1 x_1 + \omega_2 x_0) + (\omega_0 x_1 + \omega_0 x_2 + \omega_0 x_0) = 3x_2$$
$$(\omega_0 x_0 + \omega_1 x_1 + \omega_2 x_2) + (\omega_0 x_0 + \omega_1 x_2 + \omega_2 x_1) + (\omega_0 x_1 + \omega_0 x_2 + \omega_0 x_0) = 3x_0$$

The previous set of variables $\{x_1, x_2, x_3\}$ has been replaced by $\{x_1, x_2, x_0\}$ to simplify modulo 3 operations on their subscripts.

Using their exponential form, the actions performed by the *rotators* $\{\omega_0, \omega_1, \omega_2\}$ can be treated as a single static frame or as a dynamic sequence thereof.

A factor of three, rather than one, on the right side of the equal sign was chosen to avoid using fractions, rounding off, or truncation to perform left-side computations. Fractions add unnecessary complications, whereas both rounding off and truncation deteriorate the accuracy of the results. Using integers for the computations and getting the values of the individual variables by dividing the final results require less effort and provide better accuracy.

The square matrix on the left side of the equal signs is designated as a dynamic convolution of three variables (DyCon3) matrix. The adjective *dynamic* refers to the cases in which the *n* parameter in the exponential form of the rotators assumes continuously increasing (or decreasing) integer values.

Row or Column Characteristics

The DyCon3 matrix has been reproduced here to enable following the description of its features without having to flip pages.

$$(\omega_0 x_1 + \omega_1 x_2 + \omega_2 x_0) + (\omega_0 x_1 + \omega_1 x_0 + \omega_2 x_2) + (\omega_0 x_1 + \omega_0 x_2 + \omega_0 x_0) = 3x_1$$
$$(\omega_0 x_2 + \omega_1 x_0 + \omega_2 x_1) + (\omega_0 x_2 + \omega_1 x_1 + \omega_2 x_0) + (\omega_0 x_1 + \omega_0 x_2 + \omega_0 x_0) = 3x_2$$
$$(\omega_0 x_0 + \omega_1 x_1 + \omega_2 x_2) + (\omega_0 x_0 + \omega_1 x_2 + \omega_2 x_1) + (\omega_0 x_1 + \omega_0 x_2 + \omega_0 x_0) = 3x_0$$

It is a two-dimensional matrix with three rows and three columns. Each row and column contains three sets of round parentheses. Their contents are scalar products of rotator vectors $\{\omega_0, \omega_1, \omega_2\}$ and variable vectors $\{x_1, x_2, x_0\}$. The products are designated as *components*. At the end of each row is the result of performing the indicated operations on its components. The results are designated as *row identities*.

In every row, two variables from different components are organized as if they were cubic roots of the same quantity, which makes their sums equal to zero, although they cannot be removed without invalidating the row identity. The other variable has ω_0 as its rotator for all row components and for a total sum of three times its value.

In the first two columns, the order of vectors of rotators $\{\omega_0, \omega_1, \omega_2\}$ is held constant. The vectors of variables $\{x_1, x_2, x_0\}$ are circularly rotated from the top to the bottom row, toward the left for the first column and toward the right for the second column.

These first two columns are designated as *symmetrical columns*, and the trait just described is designated as the *counter rotating variables* feature.

All components in the third column are identical and use ω_0 as the only rotator.

Any row of the matrix can be rewritten in terms of any other row chosen as the *reference row* as follows:

$$\omega_0 (\omega_0 x_1 + \omega_1 x_2 + \omega_2 x_0) + \omega_0 (\omega_0 x_1 + \omega_1 x_0 + \omega_2 x_2) + \omega_0 (\omega_0 x_1 + \omega_0 x_2 + \omega_0 x_0) = 3x_1$$
$$\omega_2 (\omega_0 x_1 + \omega_1 x_2 + \omega_2 x_0) + \omega_1 (\omega_0 x_1 + \omega_1 x_0 + \omega_2 x_2) + \omega_0 (\omega_0 x_1 + \omega_0 x_2 + \omega_0 x_0) = 3x_2$$
$$\omega_1 (\omega_0 x_1 + \omega_1 x_2 + \omega_2 x_0) + \omega_2 (\omega_0 x_1 + \omega_1 x_0 + \omega_2 x_2) + \omega_0 (\omega_0 x_1 + \omega_0 x_2 + \omega_0 x_0) = 3x_0$$

$$\omega_1 (\omega_0 x_2 + \omega_1 x_0 + \omega_2 x_1) + \omega_2 (\omega_0 x_2 + \omega_1 x_1 + \omega_2 x_0) + \omega_0 (\omega_0 x_1 + \omega_0 x_2 + \omega_0 x_0) = 3x_1$$
$$\omega_0 (\omega_0 x_2 + \omega_1 x_0 + \omega_2 x_1) + \omega_0 (\omega_0 x_2 + \omega_1 x_1 + \omega_2 x_0) + \omega_0 (\omega_0 x_1 + \omega_0 x_2 + \omega_0 x_0) = 3x_2$$
$$\omega_2 (\omega_0 x_2 + \omega_1 x_0 + \omega_2 x_1) + \omega_1 (\omega_0 x_2 + \omega_1 x_1 + \omega_2 x_0) + \omega_0 (\omega_0 x_1 + \omega_0 x_2 + \omega_0 x_0) = 3x_0$$

$$\omega_2 (\omega_0 x_0 + \omega_1 x_1 + \omega_2 x_2) + \omega_1 (\omega_0 x_0 + \omega_1 x_2 + \omega_2 x_1) + \omega_0 (\omega_0 x_1 + \omega_0 x_2 + \omega_0 x_0) = 3x_1$$
$$\omega_1 (\omega_0 x_0 + \omega_1 x_1 + \omega_2 x_2) + \omega_2 (\omega_0 x_0 + \omega_1 x_2 + \omega_2 x_1) + \omega_0 (\omega_0 x_1 + \omega_0 x_2 + \omega_0 x_0) = 3x_2$$
$$\omega_0 (\omega_0 x_0 + \omega_1 x_1 + \omega_2 x_2) + \omega_0 (\omega_0 x_0 + \omega_1 x_2 + \omega_2 x_1) + \omega_0 (\omega_0 x_1 + \omega_0 x_2 + \omega_0 x_0) = 3x_0$$

The variable associated with the reference row is defined as the *reference variable*.

The first row of the original matrix is chosen as the reference row for the first new matrix; its components are rotated by ω_0 (no rotation). In the second row, the first and second components are rotated, by ω_2 and ω_1, respectively— that is, they are counter rotated. In the third row, the first and second components are rotated, by ω_1 and ω_2, respectively; they are also counter rotated but in the direction opposite that used for the second row.

In the second new matrix, the second row is chosen as the reference row; hence, all its components are rotated by ω_0 (no rotation). In the first row the first and second components are rotated, by ω_1 and by ω_2, respectively; they are counter rotated. In the third row, the first and second components are rotated, by ω_2 and by ω_1, respectively. They are counter rotated but in the direction opposite that used for the first row.

In the third new matrix, the third row is chosen as the reference row; hence, all its components are rotated by ω_0 (no rotation). In the first row, the first and second components are rotated, by ω_2 and by ω_1, respectively; they are counter rotated. In the second row, the first and second components are rotated, by ω_1 and by ω_2, respectively; they are counter rotated but in the direction opposite that used for the first row.

The new matrices show that the components in each of the first two columns are bound by a *2π/3 correspondence*; hence, they are the root elements of a 3-cluster, the power element of which is computed by raising any of their root elements to the third power. Even though one matrix is sufficient to show that the components in each of the first two columns are bound by a 2π/3 correspondence, all three matrices are used to show that the 2π/3 correspondence does not depend on which row of the original matrix was chosen as the reference row.

Together, the new matrices are designated as the *upgraded matrix triplet*.

The reference variable plays a special role within its own matrix. It is obtained by adding its row components directly as they are, whereas to get any other variable, the components of each *symmetrical column* in its row must be rotated by 2π/3 radians in opposite directions before they can be added. To get both of the other two variables, it is necessary to use both possible combinations of rotation.

The three step procedure of adding the components of the reference row as they are to obtain the reference variable; counter rotating the components of the symmetrical columns before adding them to obtain a second variable; and finally counter rotating the components of the symmetrical columns (again, but in directions opposite to those used previously) before adding them to obtain the third variables is a very important feature inherited by the descendants of the symmetrical columns.

The procedure is described for the third matrix. Its third row is the reference row. Its reference variable x_0 is obtained by adding the row components as they are. To get the first-row variable x_1, the components of the symmetrical columns are rotated, by ω_2 and ω_1, respectively, before being added. In the second row, the components of the symmetrical columns are rotated, by ω_1 and ω_2, respectively before they are added to get the variable x_2.

Computing Formulas: First Step

The power elements of the 3-clusters associated with the symmetrical columns are computed by raising to the third power any *one* of their components. Components can be chosen from different rows of any matrix in the upgraded matrix triplet. This process leads to a quadratic equation, the well-known formula of which expresses its roots as fractions. To avoid them, all the components and row identities of the first upgraded matrix are doubled as follows:

$$2[\omega_0(\omega_0 x_1 + \omega_1 x_2 + \omega_2 x_0) + \omega_0(\omega_0 x_1 + \omega_1 x_0 + \omega_2 x_2) + \omega_0(\omega_0 x_1 + \omega_0 x_2 + \omega_0 x_0)] = 6x_1$$
$$2[\omega_2(\omega_0 x_1 + \omega_1 x_2 + \omega_2 x_0) + \omega_1(\omega_0 x_1 + \omega_1 x_0 + \omega_2 x_2) + \omega_0(\omega_0 x_1 + \omega_0 x_2 + \omega_0 x_0)] = 6x_2$$
$$2[\omega_1(\omega_0 x_1 + \omega_1 x_2 + \omega_2 x_0) + \omega_2(\omega_0 x_1 + \omega_1 x_0 + \omega_2 x_2) + \omega_0(\omega_0 x_1 + \omega_0 x_2 + \omega_0 x_0)] = 6x_0$$

Designate as $C_1{}^3$ and as $C_2{}^3$ the cubic power (with the factor 2 included) of the first and second column components of the preceding matrix, respectively.

Computations are performed using related *circular groups*, such as $x_1{}^2 x_2 + x_2{}^2 x_0 + x_0{}^2 x_1$ and $x_1 x_2{}^2 + x_2 x_0{}^2 + x_0 x_1{}^2$, to emphasize that they produce symmetrical expressions.

The values of $C_1{}^3$ and $C_2{}^3$ are found as

$$(2C_1)^3 = [2\omega_0(\omega_0 x_1 + \omega_1 x_2 + \omega_2 x_0)]^3 =$$
$$= 8\{(x_1{}^3 + x_2{}^3 + x_0{}^3) + 6x_1 x_2 x_0 + 3[\omega_1(x_1{}^2 x_2 + x_2{}^2 x_0 + x_0{}^2 x_1) + \omega_2(x_1{}^2 x_0 + x_2{}^2 x_1 + x_0{}^2 x_2)]\}$$

$$(2C_2)^3 = [2\omega_0(\omega_0 x_1 + \omega_2 x_2 + \omega_1 x_0)]^3 =$$
$$= 8\{(x_1{}^3 + x_2{}^3 + x_0{}^3) + 6x_1 x_2 x_0 + 3[\omega_2(x_1{}^2 x_2 + x_2{}^2 x_0 + x_0{}^2 x_1) + \omega_1(x_1{}^2 x_0 + x_2{}^2 x_1 + x_0{}^2 x_2)]\}$$

As shown later, $C_1{}^3$ and $C_2{}^3$ have the form of quadratic roots. The quadratic could be solved by finding its coefficients as the sum and product of $C_1{}^3$ and $C_2{}^3$. Because the $C_1{}^3$ and $C_2{}^3$ expressions are symmetrical, their sum and product (as quadratic coefficients) can be expressed in terms of the COI coefficients. Known properties binding the coefficients and roots of a quadratic can then be applied to investigate the features of $C_1{}^3$ and $C_2{}^3$.

The method used here is based on a dynamic convolution of two variables (*DyCon2*) matrix. Its rotators are the square roots of unity. Instead of using their exponential form, the rotators can be written more simply as +1 and −1. Because they are used only as factors, they can be further simplified as + and −, respectively.

To avoid ambiguity, cubic roots are designated as *c-roots* and quadratic roots as *q-roots*, when needed.

DyCon2

The DyCon2 matrix of two variables *u* and *v* is written as

$(-u + v) + (u + v) = 2v$

$(+u - v) + (u + v) = 2u$

The left-column components are not symmetrical, but they are both included in the following symmetrical expressions:

$(u - v)^2 = (v - u)^2 = \Delta^2$.

Let $u + v = \Sigma$. The variables u and v can be written directly from the DyCon2 matrix as follows:

$$u, v = \frac{1}{2}\Sigma \pm \frac{1}{2}\sqrt{\Delta^2}$$

Let $u = (2C_1)^3$ and $v = (2C_2)^3$. Then using the relation $\omega_1 + \omega_2 = -1$, the sum of $(2C_1)^3$ and $(2C_2)^3$ is computed as

$\Sigma = (2C_1)^3 + (2C_2)^3 =$

$= 8[2(x_1^3 + x_2^3 + x_0^3) + 12x_1x_2x_0 - 3(x_1^2x_2 + x_2^2x_0 + x_0^2x_1 + x_1^2x_0 + x_2^2x_1 + x_0^2x_2)] =$

$= 8(-2a^3 + 9ab - 27c)$

Computations performed to change the preceding expression from a function of the COI roots to a function of its coefficients are shown in the appendix.

The midpoint between $(2C_1)^3$ and $(2C_2)^3$ is designated as the *q-root midpoint* and is given by the following

$$\tfrac{1}{2}\Sigma = \tfrac{1}{2}\lfloor (C_1)^3 + (C_2)^3 \rfloor = 4(-2a^3 + 9ab - 27c)$$

If the COI coefficients are real, then the sum of the COI quadratic roots is also real, and so is the q-root midpoint.

The differences between C_1^3 and C_2^3 are written as

$(2C_1^3 - 2C_2^3) = 8\{3(\omega_1 - \omega_2)[(x_1^2x_2 + x_2^2x_0 + x_0^2x_1) - (x_1^2x_0 + x_2^2x_1 + x_0^2x_2)]\}$

$(2C_2^3 - 2C_1^3) = 8\{3(\omega_2 - \omega_1)[(x_1^2x_2 + x_2^2x_0 + x_0^2x_1) - (x_1^2x_0 + x_2^2x_1 + x_0^2x_2)]\}$

Their squares are identical and therefore computations are shown only for the first expression, which is rewritten in a form that is easier to compute, as follows

$(2C_1^3 - 2C_2^3)^2 = (8)^2 [3(\omega_1 - \omega_2)]^2 [(x_1^2x_2 + x_2^2x_0 + x_0^2x_1) - (x_1^2x_0 + x_2^2x_1 + x_0^2x_2)]^2$

The square of the difference $(2C_1{}^3 - 2C_2{}^3)$ is repeated for convenience.

$$(2C_1{}^3 - 2C_2{}^3)^2 = (8)^2 \, [3(\omega_1 - \omega_2)]^2 \, [(x_1{}^2 x_2 + x_2{}^2 x_0 + x_0{}^2 x_1) - (x_1{}^2 x_0 + x_2{}^2 x_1 + x_0{}^2 x_2)]^2$$

Its expression $(\omega_1 - \omega_2)^2$ is computed as follows

$$(\omega_1 - \omega_2)^2 = (\omega_2 - \omega_1)^2 = \omega_1{}^2 + \omega_2{}^2 - 2\,\omega_1\omega_2 = \omega_2 + \omega_1 - 2 = -3. \text{ Therefore we get:}$$

$$(8)^2 \, [3(\omega_1 - \omega_2)]^2 = 64(-27) = -1728. \text{ Then we compute}$$

$$[(x_1{}^2 x_2 + x_2{}^2 x_0 + x_0{}^2 x_1) - (x_1{}^2 x_0 + x_2{}^2 x_1 + x_0{}^2 x_2)]^2 =$$

$$= (x_1{}^2 x_2 + x_2{}^2 x_0 + x_0{}^2 x_1)^2 + (x_1{}^2 x_0 + x_2{}^2 x_1 + x_0{}^2 x_2)^2 +$$
$$-2(x_1{}^2 x_2 + x_2{}^2 x_0 + x_0{}^2 x_1)(x_1{}^2 x_0 + x_2{}^2 x_1 + x_0{}^2 x_2) =$$

$$= (x_1{}^4 x_2{}^2 + x_2{}^4 x_0{}^2 + x_0{}^4 x_1{}^2) + 2(x_1{}^2 x_2{}^3 x_0 + x_2{}^2 x_0{}^3 x_1 + x_0{}^2 x_1{}^3 x_2) +$$
$$+(x_1{}^4 x_0{}^2 + x_2{}^4 x_1{}^2 + x_0{}^4 x_2{}^2) + 2(x_1{}^3 x_2{}^2 x_0 + x_2{}^3 x_0{}^2 x_1 + x_0{}^3 x_1{}^2 x_2) +$$
$$-2\,(x_1{}^4 x_2 x_0 + x_2{}^4 x_0 x_1 + x_0{}^4 x_1 x_2 + x_1{}^3 x_2{}^3 + x_2{}^3 x_0{}^3 + x_0{}^3 x_1{}^3 + 3 x_1{}^2 x_2{}^2 x_0{}^2).$$

It follows that

$$\Delta^2 = (2C_1{}^3 - 2C_2{}^3)^2 = -1728 \, [(x_1{}^4 x_2{}^2 + x_2{}^4 x_0{}^2 + x_0{}^4 x_1{}^2 + x_1{}^4 x_0{}^2 + x_2{}^4 x_1{}^2 + x_0{}^4 x_2{}^2) +$$
$$+ 2 x_0 x_1 x_2 (x_1 x_2{}^2 + x_2 x_0{}^2 + x_0 x_1{}^2 + x_1{}^2 x_2 + x_2{}^2 x_0 + x_0{}^2 x_1) +$$
$$-2 x_0 x_1 x_2 (x_1{}^3 + x_2{}^3 + x_0{}^3) - 6\, x_1{}^2 x_2{}^2 x_0{}^2 - 2(x_1{}^3 x_2{}^3 + x_2{}^3 x_0{}^3 + x_0{}^3 x_1{}^3)] =$$

$$= -1728 \, (a^2 b^2 - 4a^3 c - 4b^3 + 18abc - 27c^2)$$

Details of the computations used to translate the preceding expression from a function of the COI roots to a function of its coefficients are also shown in the appendix.

The quadratic with $(2C_1)^3$ and $(2C_2)^3$ as its roots is designated as the *COI quadratic*. Its formula is written as follows:

$$u, v = \frac{1}{2}\Sigma \pm \frac{1}{2}\sqrt{\Delta^2} = 4(-2a^3 + 9ab - 27c) \pm \frac{1}{2}\sqrt{-1728(a^2 b^2 - 4a^3 c - 4b^3 + 18abc - 27c^2)}$$

It is interesting to note that although $(2C_1)^3$ and $(2C_2)^3$ are computed as roots of the same quadratic, they are associated with a different symmetrical column of the upgraded matrices and appear as terms in the radicands of different radicals in the cubic formula.

The DyCon3 matrix used to search for a COI formula is rewritten as follows:

$$2[\omega_0 \, (\omega_0 x_1 + \omega_1 x_2 + \omega_2 x_0) + \omega_0 \, (\omega_0 x_1 + \omega_1 x_0 + \omega_2 x_2) + \omega_0 \, (\omega_0 x_1 + \omega_0 x_2 + \omega_0 x_0)] = 6x_1$$
$$2[\omega_2 \, (\omega_0 x_1 + \omega_1 x_2 + \omega_2 x_0) + \omega_1 \, (\omega_0 x_1 + \omega_1 x_0 + \omega_2 x_2) + \omega_0 \, (\omega_0 x_1 + \omega_0 x_2 + \omega_0 x_0)] = 6x_2$$
$$2[\omega_1 \, (\omega_0 x_1 + \omega_1 x_2 + \omega_2 x_0) + \omega_2 \, (\omega_0 x_1 + \omega_1 x_0 + \omega_2 x_2) + \omega_0 \, (\omega_0 x_1 + \omega_0 x_2 + \omega_0 x_0)] = 6x_0$$

Let y be a generic root of the cubic formula based on the DyCon3 matrix and on the COI quadratic formula. An initial symbolic formula for the COI is

$$6y = \sqrt[3]{\frac{(C_1^3 + C_2^3)}{2} + \sqrt{\frac{(C_1^3 - C_2^3)^2}{4}}} + \sqrt[3]{\frac{(C_1^3 + C_2^3)}{2} - \sqrt{\frac{(C_1^3 - C_2^3)^2}{4}}} - 2a$$

On rearranging its terms and as a function of its coefficients, the COI formula becomes

$$6y = -2a + \sqrt[3]{4(-2a^3 + 9ab - 27c) + \sqrt{-432(a^2b^2 - 4a^3c - 4b^3 + 18abc - 27c^2)}}$$
$$+ \sqrt[3]{4(-2a^3 + 9ab - 27c) - \sqrt{-432(a^2b^2 - 4a^3c - 4b^3 + 18abc - 27c^2)}}$$

The classic cubic formula is derived from one that was published in the book *Artis magnae, sive de regulis algebraicis* (also known as *Ars Magna*) by Girolamo Cardano in 1545. In modern notation, it can be written as

$$y = -\frac{1}{3}a + \frac{1}{3}\sqrt[3]{\frac{-2a^3 + 9ab - 27c}{2} + \sqrt{\frac{-27(a^2b^2 - 4a^3c - 4b^3 + 18abc - 27c^2)}{4}}}$$
$$+ \frac{1}{3}\sqrt[3]{\frac{-2a^3 + 9ab - 27c}{2} - \sqrt{\frac{-27(a^2b^2 - 4a^3c - 4b^3 + 18abc - 27c^2)}{4}}}$$

The new and the classic formulas are essentially identical. Both contain three terms. Two are cubic radicals, and each generates three roots for a total of nine combinations for the variable y. Because a cubic equation can only have three roots, to fully solve the problem it is necessary to determine the reasons why the formula generates the other six combinations and how they relate to the roots of the COI.

The old formula was found via polynomial manipulations of the equation coefficients. Symmetry of the roots was not considered; probably, it was unknown. The discovery of the ESFs of the roots is attributed to François Viète (1540–1603), whereas the original formula was published in 1545. The formula was intended to solve practical problems, and the cubic coefficients were limited to positive real numbers. As a result, at least one root had to be a real number—a fact that facilitates finding a first real root for the equation. The other two roots followed by counter rotating and then adding two terms of the real root but without explaining why. Literature searches to determine whether any attempts were ever made to explain the presence of the extra six combinations proved unsuccessful. It appears that once the desired solution had been found, the fact that it could be a part of a larger problem did not seem worth pursuing.

The new formula was derived by deliberately assembling the roots as a matrix that exploits their symmetrical properties. All computations were performed in terms of the COI roots, and they all produced symmetrical expressions. A generic procedure specifically designed for that purpose was then used to translate the symmetrical functions of the roots into functions of the COI ESFs and subsequently to functions of the COI coefficients. The new formula also has two cubic radicals, which were derived from the symmetrical columns of the upgraded matrix triplet. Interchanging the first and second columns would not affect any of the matrix basic properties; the column power elements would be interchanged, as would the formula's radicals.

The third column of the matrix is the same for all rows and is related to the COI *linear coefficient a.* It normally appears as the first term in the cubic formula.

While searching for clues about why the formula generates the other six combinations and how they relate to the roots of the COI, I noticed that the quantities C_1^3 and C_2^3 do not depend on how the components of the symmetrical columns are aligned to form matrix rows. They are not affected if one column is rotated with respect to the other, although all row identities would no longer be valid.

The effects of these rotations are now scrutinized. The first of the three upgraded matrices is rewritten first as is, and then a color scheme is added to facilitate comparing its features with those of two new matrices. Recall that the chosen upgraded matrix was derived from the original matrix by selecting its first row as the matrix reference row; hence, x_1 plays the major role. Had the second row been chosen, x_2 would play the major role, and had the third row been chosen, x_0 would play the major role.

$$2[\omega_0(\omega_0 x_1 + \omega_1 x_2 + \omega_2 x_0) + \omega_0(\omega_0 x_1 + \omega_2 x_2 + \omega_1 x_0) + \omega_0(\omega_0 x_1 + \omega_0 x_2 + \omega_0 x_0)] = 6\omega_0 x_1$$
$$2[\omega_2(\omega_0 x_1 + \omega_1 x_2 + \omega_2 x_0) + \omega_1(\omega_0 x_1 + \omega_2 x_2 + \omega_1 x_0) + \omega_0(\omega_0 x_1 + \omega_0 x_2 + \omega_0 x_0)] = 6\omega_0 x_2$$
$$2[\omega_1(\omega_0 x_1 + \omega_1 x_2 + \omega_2 x_0) + \omega_2(\omega_0 x_1 + \omega_2 x_2 + \omega_1 x_0) + \omega_0(\omega_0 x_1 + \omega_0 x_2 + \omega_0 x_0)] = 6\omega_0 x_0$$

$$2[\omega_0(\omega_0 x_1 + \omega_1 x_2 + \omega_2 x_0) + \omega_1(\omega_0 x_1 + \omega_2 x_2 + \omega_1 x_0) + \omega_2(\omega_0 x_1 + \omega_0 x_2 + \omega_0 x_0)] = 6\omega_2 x_0$$
$$2[\omega_2(\omega_0 x_1 + \omega_1 x_2 + \omega_2 x_0) + \omega_2(\omega_0 x_1 + \omega_2 x_2 + \omega_1 x_0) + \omega_2(\omega_0 x_1 + \omega_0 x_2 + \omega_0 x_0)] = 6\omega_2 x_1$$
$$2[\omega_1(\omega_0 x_1 + \omega_1 x_2 + \omega_2 x_0) + \omega_0(\omega_0 x_1 + \omega_2 x_2 + \omega_1 x_0) + \omega_2(\omega_0 x_1 + \omega_0 x_2 + \omega_0 x_0)] = 6\omega_2 x_2$$

$$2[\omega_0(\omega_0 x_1 + \omega_1 x_2 + \omega_2 x_0) + \omega_2(\omega_0 x_1 + \omega_2 x_2 + \omega_1 x_0) + \omega_1(\omega_0 x_1 + \omega_0 x_2 + \omega_0 x_0)] = 6\omega_1 x_2$$
$$2[\omega_2(\omega_0 x_1 + \omega_1 x_2 + \omega_2 x_0) + \omega_0(\omega_0 x_1 + \omega_2 x_2 + \omega_1 x_0) + \omega_1(\omega_0 x_1 + \omega_0 x_2 + \omega_0 x_0)] = 6\omega_1 x_0$$
$$2[\omega_1(\omega_0 x_1 + \omega_1 x_2 + \omega_2 x_0) + \omega_1(\omega_0 x_1 + \omega_2 x_2 + \omega_1 x_0) + \omega_1(\omega_0 x_1 + \omega_0 x_2 + \omega_0 x_0)] = 6\omega_1 x_1$$

To get the second matrix, the second column of the first matrix is circulated by one step, which moves its top row to the bottom. To get the third matrix, the second column of the first matrix is circulated by one step, which moves its bottom row to the top. To reestablish the validity of the row operations, the third-column component and the row identities have to be modified.

Note that the second-column components of the first matrix are rotated in one direction to obtain the second matrix and in the opposite direction to obtain the third matrix. As mentioned earlier, instances of rotations occurring in pairs, in opposite directions and together with a no rotation seem to be an intrinsic feature of cubic symmetrical structures, and they occur frequently.

Together, the matrices above are designated as the *COI triplet of matrices* or *COI triplet* for short. The arrangement of the COI roots and their associated constituent terms as defined by the first matrix is designated as the *COI structure.*

The COI triplet main features and their consequences are as follows:

1. Each matrix is associated with a cubic.
2. The components in the same columns of the COI triplet are identical.
3. The radicals of the initial cubic formula were generated using only the symmetrical columns of the first matrix. Therefore, all three matrices will generate identical radicals for their formulas.
4. The second matrix is obtained also multiplying every row of the first matrix and its row identity by ω_2. This affects only the row identity variable; the other variables retain their zero value. Hence, rotating by ω_2 all the parts that define the COI structure generates all the parts of a second cubic.
5. Similarly, the third matrix is obtained multiplying every row of the first matrix and its row identity by ω_1. Again, this affects only the row identity variable; the other variables retain their zero value. Hence, rotating by ω_1 all the parts that define the COI structure generates all the parts of a third cubic.
6. Combining four and five above can be stated as, "For every part of the COI structure, there is a correspondent one located $2\pi/3$ apart in either direction." In other words, the roots of the three cubics in the triplet are bound by a $2\pi/3$ correspondence.
7. In particular, the third columns of the COI triplet form a symmetric structure and provide the missing correspondent parts for the first term of the initial cubic formula. Together, they form a 3-cluster, the root elements of which can be written as

$$-2a\sqrt[3]{1} = \{-2a, -2a\omega_1, -2a\omega_2\}$$

This 3-cluster is designated as the *L-cluster* (linear cluster) and its roots as *L-roots*. As it will be shown later, is center is also the center of the overall cubic structure.

There are similarities and differences between the features of the COI triplet of matrices and those of the upgraded matrix triplet (pages 10 and 11).

The matrices of the upgraded matrix triplet are a different form of the same DyCon3 matrix, and together, they indicate that the components in each of its symmetrical columns are the root elements of the same 3-cluster. Their power element is obtained as the cubic power of any root element located in the same column of any matrix. All three matrices are associated with the COI.

In the COI triplet of matrices, each matrix is the DyCon3 matrix of a different cubic. Their forms indicate that their roots are bound by a $2\pi/3$ correspondence. Any cubic can be used as a reference for the other two, and if used as the new COI, it generates the same cubic formula and the same symmetrical structure.

The COI triplet matrices have features that are useful for computing their right-side variables (that is, their cubic roots). Every matrix has a reference row that is associated with x_1, and it is characterized by having all its row items rotated by the same rotator. For the first matrix, they are the first row, with ω_0 as the rotator; for the second matrix, they are the second row with ω_2 as the rotator; and for the third matrix, they are the third row with ω_1 as the rotator.

The rules for computing the cubic root associated with any row of any matrix are as follows:

1. Locate the matrix reference row.
2. Add its components as they are to get the reference root.
3. Group the two non-reference rotators as two differently ordered pairs.
4. Use the first pair of rotators to get a second cubic root via the following two-step procedure:
   a.   Rotate the first component of the reference row by the first rotator and the second component of the reference row by the second rotator.
   b.   Add the rotated components to the common L-root of the matrix.
5. Use the second pair of rotators to get the third cubic root via the two-step procedure above.

The cubic formula can be upgraded to the following, more symmetrical expression:

$$6y = -2a\sqrt[3]{1} + \sqrt[3]{4(-2a^3 + 9ab - 27c) + \sqrt{-432(a^2b^2 - 4a^3c - 4b^3 + 18abc - 27c^2)}}$$
$$+ \sqrt[3]{4(-2a^3 + 9ab - 27c) - \sqrt{-432(a^2b^2 - 4a^3c - 4b^3 + 18abc - 27c^2)}}$$

Each term of the formula generates three roots for a total of twenty-seven possible solutions, whereas the matrix triplet can only account for nine. The new formula seems to have uncovered even more parts without correspondent ones than did the old formula.

Because the formula radicands have the format of a pair of quadratic roots, the features of a generic quadratic and of its roots are now scrutinized to search for more clues on how to create fully symmetrical formulas for both the quadratic and the cubic equations.

Quadratic: Formulas and Features

The *COI* quadratic formula has faults that need to be addressed. The formula is repeated below for convenience.

$$u, v = \frac{1}{2}\Sigma \pm \frac{1}{2}\sqrt{\Delta^2} = 4(-2a^3 + 9ab - 27c) \pm \sqrt{-432(a^2b^2 - 4a^3c - 4b^3 + 18abc - 27c^2)}$$

It contains a first expression that does not have a symmetrical correspondent.

Designate as the COI *opposite* cubic a new cubic with roots $\{-x_1, -x_2, -x_0\}$—that is, one such that its roots form additive inverses with the COI root with equal subscript.

Let $\{a_1, b_1, c_1\}$ be the coefficients of the COI *opposite* cubic. The formula of its associated quadratic (designated as the COI *opposite quadratic*) is written as

$$u, v = 4(-2a_1^3 + 9a_1b_1 - 27c_1) \pm \sqrt{-432(a_1^2b_1^2 - 4a_1^3c_1 - 4b_1^3 + 18a_1b_1c_1 - 27c_1^3)}$$

The COI *opposite* cubic coefficients $\{a_1, b_1, c_1\}$ and some of their relevant products are related to those of the COI coefficients $\{a, b, c\}$ as follows:

$$a_1 = -a; \; b_1 = b; \; c_1 = -c; \; a_1^3 = -a^3; \; a_1 b_1 = -a\,b$$

$$a_1^2 b_1^2 = a^2 b^2; \; a_1^3 c_1 = a^3 c; \; b_1^3 = b^3; \; a_1 b_1 c_1 = abc; \; c_1^2 = c^2$$

In terms of the COI coefficients, the COI opposite quadratic first expression is written as

$$-4(-2a^3 + 9ab - 27c)$$

Therefore, the first expressions of the COI quadratic and that of the COI opposite quadratic are additive inverses and together can be written as

$$4(-2a^3 + 9ab - 27c)\sqrt{1}$$

Notice that the first expressions in the formulas of the COI quadratic and of the COI opposite quadratic must have different signs; formulas in which the signs are equal are disallowed because their roots are not roots of either quadratic.

A *symmetrical quadratic* formula, including both the *opposite* quadratic and the *COI quadratic*, could be written as

$$u, v = \frac{1}{2}\Sigma \pm \frac{1}{2}\sqrt{\Delta^2} = 4(-2a^3 + 9ab - 27c)\sqrt{1} \pm \sqrt{-432(a^2 b^2 - 4a^3 c - 4b^3 + 18abc - 27c^2)}$$

The value of the square root of unity (+1 or −1) must be held constant for all the cubics in the COI triplet and for all the cubics in the *opposite-matrix* triplet but it must be different for each triplet. This condition will be referred to as the *COI opposite quadratics' restriction*. This topic will be discussed again from page 37 to 38 and again from page 49 to 50.

The upgraded cubic COI *formula* was shared by the COI triplet of cubics and is shown on the opposite page. Similarly, the formula of the COI *opposite cubic* is written as

$$6y = -2a\sqrt[3]{1} + \sqrt[3]{-4(-2a^3 + 9ab - 27c) + \sqrt{-432(a^2 b^2 - 4a^3 c - 4b^3 + 18abc - 27c^2)}}$$
$$+ \sqrt[3]{-4(-2a^3 + 9ab - 27c) - \sqrt{-432(a^2 b^2 - 4a^3 c - 4b^3 + 18abc - 27c^2)}}$$

The two formulas could be combined as follows:

$$6y = -2a\sqrt[3]{1} + \sqrt[3]{4(-2a^3 + 9ab - 27c)\sqrt{1} + \sqrt{-432(a^2 b^2 - 4a^3 c - 4b^3 + 18abc - 27c^2)}}$$
$$+ \sqrt[3]{4(-2a^3 + 9ab - 27c)\sqrt{1} - \sqrt{-432(a^2 b^2 - 4a^3 c - 4b^3 + 18abc - 27c^2)}}$$

However, the formula does not clearly convey the COI opposite quadratics' restriction. This topic will be revisited later.

It is interesting to note that the rules of symmetry lead to the conclusion that to get the *COI fully symmetrical structure*, two separate but symmetrically related formulas may be needed. The formulas would be solved separately and the resulting structures merged into a single structure.

Quadratic Formula: More useful Features

The features of a quadratic and those of its roots are scrutinized to better understand the role they play in the cubic formula. The quadratic equation and its roots $\{z_1, z_2\}$ are normally written as

$$az^2 + bz + c = 0$$

$$z_1, z_2 = \frac{1}{2a}\left(-b \pm \sqrt{b^2 - 4ac}\right)$$

The quadratic a coefficient cannot be zero; the formula would be meaningless and the equation would not be a quadratic. Let

$$b_1 = \frac{b}{a} \quad and \quad c_1 = \frac{c}{a}$$

The coefficients of the equation, the formula, its roots, their sum, and their product can be rewritten as

$$z^2 + b_1 z + c_1 = 0$$

$$z_1, z_2 = \frac{1}{2}\left(-b_1 \pm \sqrt{b_1^2 - 4c_1}\right)$$

$$z_1 + z_2 = -b_1 = \Sigma; \qquad z_1 z_2 = \frac{1}{4}(b_1^2 - b_1^2 + 4c_1) = c_1$$

The radicand can be written in terms of the square of the root differences as follows:

$$b_1^2 - 4c_1 = z_1^2 + z_2^2 + 2z_1 z_2 - 4z_1 z_2 = (z_1 - z_2)^2 = (z_2 - z_1)^2 = \Delta^2$$

The coefficients b_1 and c_1 are arbitrarily given real numbers. As a function of the sum and product of the roots, the radicand can assume negative values but not as a function of the square of the root differences if root values are limited to real numbers.

The roots are symmetrically located about their midpoint at a distance that is equal to one-half the magnitudes of their differences, as indicated next:

$$z_1 = -\frac{1}{2}\Sigma + \frac{1}{2}\Delta \; ; \quad z_2 = -\frac{1}{2}\Sigma - \frac{1}{2}\Delta$$

This type of symmetry is designated as *local symmetry*, and it is graphically depicted in figure 2, which employs a set of Cartesian coordinates with a common imaginary axis (solid, black vertical line) and four real axes (dotted, black, horizontal lines).

The vertical, dashed line labeled as $\frac{1}{2}\Sigma$ indicates the midpoint of the roots. Its location is held constant, and its value is assumed to be positive. The roots and their midpoints are assumed to be real. Their locations are chosen in a sequence that shows how the formula radicand behaves as its value becomes progressively smaller.

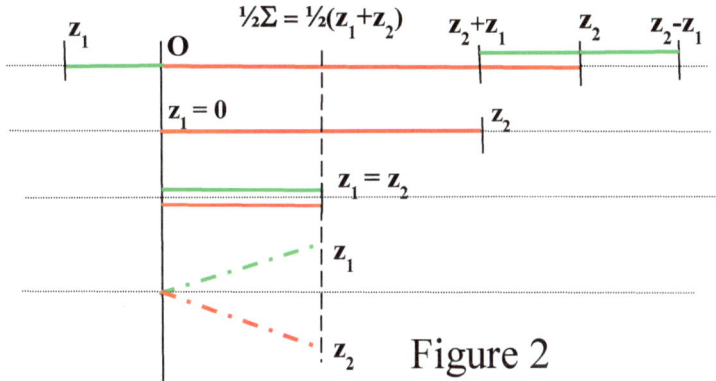

Figure 2

The roots on the top real axis have opposite signs. Their product is negative; the formula radicand is positive and larger than the square of the sum of the roots.

One root on the next axis is equal to zero, and the other is positive. Their product is zero; the formula radicand is equal to the square of the sum of the roots. It is positive, but it has a value smaller than the previous one.

On the next axis, the roots are equal and positive; their difference is equal to zero, and the formula radicand is zero. More significant is the fact that the product of the roots has reached its maximum value because the product of two real numbers with a constant sum reaches its maximum value when the numbers are equal. Under the assumption that the roots are real, the radicand has reached its minimum value of zero.

The square of any real number is always positive. A negative radicand (the result of $b_1^2 < 4c_1$) is made positive by rotating it 180°. The task can also be accomplished by first rotating its basis (the square root of the radicand) by 90°.

The symbol j (or i) is normally used as a prefix for imaginary numbers. If treated as a factor, the prefix works as an algebraic operator that rotates any associated quantity by 90° — that is, it works as a *90° rotator*.

The quadratic roots for a negative radicand are depicted at the bottom real axis of figure 2. Note that the midpoint of the roots has not changed, and it still acts as the local symmetry center for the half differences of the roots that are now shown as vertical,

21

rather than horizontal segments.

In reality, the root differences are and should be treated as having been moved into the second dimension of a two-dimensional coordinate system. Using imaginary numbers is just a way of adding a second dimension.

The earlier assumption that the root midpoint be a positive quantity is not necessary. If the root midpoint happens to be negative, figure 2 would change into its own mirror image with respect to the imaginary axis. By appropriately changing some phrasing, the process and its description would still be valid and so would its conclusions.

Matrix Sextuplet

Matrices with components that differ only by their algebraic sign are designated as *opposite matrices*. Listed below are the COI triplet of matrices and the matrix triplet associated with the COI *opposite* cubic. The second triplet is designated as the *opposite-matrix triplet*. For instance, the first matrix is the COI matrix, the fourth matrix is the COI *opposite*-cubic matrix.

Together, the triplets are designated as a *matrix sextuplet*.

$$2\omega_0(\omega_0 x_1 + \omega_1 x_2 + \omega_2 x_0) + \omega_0(\omega_0 x_1 + \omega_2 x_2 + \omega_1 x_0) + \omega_0(\omega_0 x_1 + \omega_0 x_2 + \omega_0 x_0) = 6\omega_0 x_1$$
$$2\omega_2(\omega_0 x_1 + \omega_1 x_2 + \omega_2 x_0) + \omega_1(\omega_0 x_1 + \omega_2 x_2 + \omega_1 x_0) + \omega_0(\omega_0 x_1 + \omega_0 x_2 + \omega_0 x_0) = 6\omega_0 x_2$$
$$2\omega_1(\omega_0 x_1 + \omega_1 x_2 + \omega_2 x_0) + \omega_2(\omega_0 x_1 + \omega_2 x_2 + \omega_1 x_0) + \omega_0(\omega_0 x_1 + \omega_0 x_2 + \omega_0 x_0) = 6\omega_0 x_0$$

$$2\omega_0(\omega_0 x_1 + \omega_1 x_2 + \omega_2 x_0) + \omega_1(\omega_0 x_1 + \omega_2 x_2 + \omega_1 x_0) + \omega_2(\omega_0 x_1 + \omega_0 x_2 + \omega_0 x_0) = 6\omega_2 x_0$$
$$2\omega_2(\omega_0 x_1 + \omega_1 x_2 + \omega_2 x_0) + \omega_2(\omega_0 x_1 + \omega_2 x_2 + \omega_1 x_0) + \omega_2(\omega_0 x_1 + \omega_0 x_2 + \omega_0 x_0) = 6\omega_2 x_1$$
$$2\omega_1(\omega_0 x_1 + \omega_1 x_2 + \omega_2 x_0) + \omega_0(\omega_0 x_1 + \omega_2 x_2 + \omega_1 x_0) + \omega_2(\omega_0 x_1 + \omega_0 x_2 + \omega_0 x_0) = 6\omega_2 x_2$$

$$2\omega_0(\omega_0 x_1 + \omega_1 x_2 + \omega_2 x_0) + \omega_2(\omega_0 x_1 + \omega_2 x_2 + \omega_1 x_0) + \omega_1(\omega_0 x_1 + \omega_0 x_2 + \omega_0 x_0)] = 6\omega_1 x_2$$
$$2\omega_2(\omega_0 x_1 + \omega_1 x_2 + \omega_2 x_0) + \omega_0(\omega_0 x_1 + \omega_2 x_2 + \omega_1 x_0) + \omega_1(\omega_0 x_1 + \omega_0 x_2 + \omega_0 x_0)] = 6\omega_1 x_0$$
$$2\omega_1(\omega_0 x_1 + \omega_1 x_2 + \omega_2 x_0) + \omega_1(\omega_0 x_1 + \omega_2 x_2 + \omega_1 x_0) + \omega_1(\omega_0 x_1 + \omega_0 x_2 + \omega_0 x_0)] = 6\omega_1 x_1$$

$$2\omega_0(\omega_0 \underline{x_1} + \omega_1 \underline{x_2} + \omega_2 \underline{x_0}) + \omega_0(\omega_0 \underline{x_1} + \omega_2 \underline{x_2} + \omega_1 \underline{x_0}) + \omega_0(\omega_0 \underline{x_1} + \omega_0 \underline{x_2} + \omega_0 \underline{x_0}) = 6\omega_0 \underline{x_1}$$
$$2\omega_2(\omega_0 \underline{x_1} + \omega_1 \underline{x_2} + \omega_2 \underline{x_0}) + \omega_1(\omega_0 \underline{x_1} + \omega_2 \underline{x_2} + \omega_1 \underline{x_0}) + \omega_0(\omega_0 \underline{x_1} + \omega_0 \underline{x_2} + \omega_0 \underline{x_0}) = 6\omega_0 \underline{x_2}$$
$$2\omega_1(\omega_0 \underline{x_1} + \omega_1 \underline{x_2} + \omega_2 \underline{x_0}) + \omega_2(\omega_0 \underline{x_1} + \omega_2 \underline{x_2} + \omega_1 \underline{x_0}) + \omega_0(\omega_0 \underline{x_1} + \omega_0 \underline{x_2} + \omega_0 \underline{x_0}) = 6\omega_0 \underline{x_0}$$

$$2\omega_0(\omega_0 \underline{x_1} + \omega_1 \underline{x_2} + \omega_2 \underline{x_0}) + \omega_1(\omega_0 \underline{x_1} + \omega_2 \underline{x_2} + \omega_1 \underline{x_0}) + \omega_2(\omega_0 \underline{x_1} + \omega_0 \underline{x_2} + \omega_0 \underline{x_0}) = 6\omega_2 \underline{x_0}$$
$$2\omega_2(\omega_0 \underline{x_1} + \omega_1 \underline{x_2} + \omega_2 \underline{x_0}) + \omega_2(\omega_0 \underline{x_1} + \omega_2 \underline{x_2} + \omega_1 \underline{x_0}) + \omega_2(\omega_0 \underline{x_1} + \omega_0 \underline{x_2} + \omega_0 \underline{x_0}) = 6\omega_2 \underline{x_1}$$
$$2\omega_1(\omega_0 \underline{x_1} + \omega_1 \underline{x_2} + \omega_2 \underline{x_0}) + \omega_0(\omega_0 \underline{x_1} + \omega_2 \underline{x_2} + \omega_1 \underline{x_0}) + \omega_2(\omega_0 \underline{x_1} + \omega_0 \underline{x_2} + \omega_0 \underline{x_0}) = 6\omega_2 \underline{x_2}$$

$$2\omega_0(\omega_0 \underline{x_1} + \omega_1 \underline{x_2} + \omega_2 \underline{x_0}) + \omega_2(\omega_0 \underline{x_1} + \omega_2 \underline{x_2} + \omega_1 \underline{x_0}) + \omega_1(\omega_0 \underline{x_1} + \omega_0 \underline{x_2} + \omega_0 \underline{x_0}) = 6\omega_1 \underline{x_2}$$
$$2\omega_2(\omega_0 \underline{x_1} + \omega_1 \underline{x_2} + \omega_2 \underline{x_0}) + \omega_0(\omega_0 \underline{x_1} + \omega_2 \underline{x_2} + \omega_1 \underline{x_0}) + \omega_1(\omega_0 \underline{x_1} + \omega_0 \underline{x_2} + \omega_0 \underline{x_0}) = 6\omega_1 \underline{x_0}$$
$$2\omega_1(\omega_0 \underline{x_1} + \omega_1 \underline{x_2} + \omega_2 \underline{x_0}) + \omega_1(\omega_0 \underline{x_1} + \omega_2 \underline{x_2} + \omega_1 \underline{x_0}) + \omega_1(\omega_0 \underline{x_1} + \omega_0 \underline{x_2} + \omega_0 \underline{x_0}) = 6\omega_1 \underline{x_1}$$

Variable underlining is used in the preceding sextuplet as a means to indicate that the underlined variable has a negative sign, that is: $\underline{x_1}$ = -x_1, $\underline{x_2}$ = - x_2, and $\underline{x_0}$ = - x_0.

The components of each symmetrical column are identical for matrices in the same triplet, and they are additive inverses for those in different triplets.

A more efficient way for describing and using the features of the matrix sextuplet is to perform the following simplifications.

1. Avoid using negative signs by replacing the COI coefficient a with the ESF σ_1.
2. Divide both sides of all equalities by two.
3. Replace the symmetrical components with a shorter notation.
4. Replace the $2\pi/3$ correspondence with a $\pi/3$ correspondence.

To satisfy three above, again let

$$C_1 = \omega_0 x_1 + \omega_1 x_2 + \omega_2 x_0; \quad C_2 = \omega_0 x_1 + \omega_2 x_2 + \omega_1 x_0$$

To satisfy both 1 and 4, use the principal roots of $\sqrt[6]{1}$. These are defined as {δ_1, δ_2, δ_3, δ_4, δ_5, δ_6}. Their angles are {60°, 120°, 180°, 240°, 300°, 0°}. These are designated as the δ rotators and are related to those of the {ω_0, ω_1, ω_2} set as follows:

$$\partial_1 = -\omega_2; \quad \partial_2 = \omega_1; \quad \partial_3 = -\omega_0; \quad \partial_4 = \omega_2; \quad \partial_5 = -\omega_1; \quad \partial_0 = \omega_0 = 1$$

Rotator pairs {δ_0, δ_3}, {δ_1, δ_4}, and {δ_2, δ_5} are diametrically opposite. The difference between the subscripts of each pair is equal to zero modulo 3. As a result, the septuplet can be written as

$[\delta_0 C_1 + \delta_0 C_2 + \delta_0 \sigma_1] = 3\delta_0 x_1$

$[\delta_4 C_1 + \delta_2 C_2 + \delta_0 \sigma_1] = 3\delta_0 x_2$

$[\delta_2 C_1 + \delta_4 C_2 + \delta_0 \sigma_1] = 3\delta_0 x_0$

$[\delta_0 C_1 + \delta_2 C_2 + \delta_4 \sigma_1] = 3\delta_4 x_0$

$[\delta_4 C_1 + \delta_4 C_2 + \delta_4 \sigma_1] = 3\delta_4 x_1$

$[\delta_2 C_1 + \delta_0 C_2 + \delta_4 \sigma_1] = 3\delta_4 x_2$

$[\delta_0 C_1 + \delta_4 C_2 + \delta_2 \sigma_1] = 3\delta_2 x_2$

$[\delta_4 C_1 + \delta_0 C_2 + \delta_2 \sigma_1] = 3\delta_2 x_0$

$[\delta_2 C_1 + \delta_2 C_2 + \delta_2 \sigma_1] = 3\delta_2 x_1$

$[\delta_3 C_1 + \delta_3 C_2 + \delta_3 \sigma_1] = 3 \delta_3 x_1$

$[\delta_1 C_1 + \delta_5 C_2 + \delta_3 \sigma_1] = 3 \delta_3 x_2$

$[\delta_5 C_1 + \delta_1 C_2 + \delta_3 \sigma_1] = 3 \delta_3 x_0$

$[\delta_3 C_1 + \delta_5 C_2 + \delta_1 \sigma_1] = 3 \delta_1 x_0$

$[\delta_1 C_1 + \delta_1 C_2 + \delta_1 \sigma_1] = 3 \delta_1 x_1$

$[\delta_5 C_1 + \delta_3 C_2 + \delta_1 \sigma_1] = 3 \delta_1 x_2$

$[\delta_3 C_1 + \delta_1 C_2 + \delta_5 \sigma_1] = 3 \delta_5 x_2$

$[\delta_1 C_1 + \delta_3 C_2 + \delta_5 \sigma_1] = 3 \delta_5 x_0$

$[\delta_5 C_1 + \delta_5 C_2 + \delta_5 \sigma_1] = 3 \delta_5 x_1$

The above sextuplet will be referred to as the COI *upgraded matrix sextuplet*.

$$[\delta_0 C_1 + \delta_0 C_2 + \delta_0 \sigma_1] = 3\delta_0 x_1 \qquad [\delta_3 C_1 + \delta_3 C_2 + \delta_3 \sigma_1] = 3\,\delta_3 x_1$$
$$[\delta_4 C_1 + \delta_2 C_2 + \delta_0 \sigma_1] = 3\delta_0 x_2 \qquad [\delta_1 C_1 + \delta_5 C_2 + \delta_3 \sigma_1] = 3\,\delta_3 x_2$$
$$[\delta_2 C_1 + \delta_4 C_2 + \delta_0 \sigma_1] = 3\delta_0 x_0 \qquad [\delta_5 C_1 + \delta_1 C_2 + \delta_3 \sigma_1] = 3\,\delta_3 x_0$$

$$[\delta_0 C_1 + \delta_2 C_2 + \delta_4 \sigma_1] = 3\delta_4 x_0 \qquad [\delta_3 C_1 + \delta_5 C_2 + \delta_1 \sigma_1] = 3\,\delta_1 x_0$$
$$[\delta_4 C_1 + \delta_4 C_2 + \delta_4 \sigma_1] = 3\delta_4 x_1 \qquad [\delta_1 C_1 + \delta_1 C_2 + \delta_1 \sigma_1] = 3\,\delta_1 x_1$$
$$[\delta_2 C_1 + \delta_0 C_2 + \delta_4 \sigma_1] = 3\delta_4 x_2 \qquad [\delta_5 C_1 + \delta_3 C_2 + \delta_1 \sigma_1] = 3\,\delta_1 x_2$$

$$[\delta_0 C_1 + \delta_4 C_2 + \delta_2 \sigma_1] = 3\delta_2 x_2 \qquad [\delta_3 C_1 + \delta_1 C_2 + \delta_5 \sigma_1] = 3\,\delta_5 x_2$$
$$[\delta_4 C_1 + \delta_0 C_2 + \delta_2 \sigma_1] = 3\delta_2 x_0 \qquad [\delta_1 C_1 + \delta_3 C_2 + \delta_5 \sigma_1] = 3\,\delta_5 x_0$$
$$[\delta_2 C_1 + \delta_2 C_2 + \delta_2 \sigma_1] = 3\delta_2 x_1 \qquad [\delta_5 C_1 + \delta_5 C_2 + \delta_5 \sigma_1] = 3\,\delta_5 x_1$$

The preceding matrix sextuplet is repeated above as a convenience. It is shown in a compact, two-column format. The left column is derived from the COI triplet, and the right column is derived from the COI opposite-matrix triplet.

A review of their salient characteristics follows:

1. The (C_1, C_2, σ_1) components are located in the same column and are identical for all matrices.
2. All *rotated* components located on the same row but from different triplets form additive inverses.
3. All components are multiplied by a rotator as a means to indicate a change of direction. In particular, ω_0 and δ_0 indicate no change; δ_3 indicates a 180° rotation.
4. Every matrix is characterized by a reference row and its reference rotator. Their definitions and uses are reviewed and expanded as follows:
   a. The reference row is defined as the row for which the rotators of all its components and of its identity (its c-root) have the same value. For all triplets, these rows are associated with the c-root x_1 or its additive inverse $-x_1$. This is not due to a special feature of the c-root x_1, but it is a consequence of having selected the first row of the original COI DyCon3 matrix as the global-reference row. For both triplets, the reference rows are the first row of the first matrix, the second row of the second matrix, and the third row of the third matrix.
   b. The reference rotator is defined as the common rotator of the reference row. For the first, second, and third matrices, they are respectively $\{\omega_0, \omega_2, \omega_1\}$ (or $\{\delta_0, \delta_4, \delta_2\}$) for the COI triplet and $\{-\omega_0, -\omega_2, -\omega_1\}$ (or $\{\delta_3, \delta_1, \delta_5\}$) for the COI *opposite* triplet.

The rules for computing the c-root of any row of a matrix are designated as *DyCon3 addition rules*. These have been explained before but are repeated now, as they apply specifically to the preceding matrices.

Every matrix has a reference row that is characterized by having the same rotator for all its components and for its identity (which is related to the c-root x_1). The rotator is designated as the reference rotator. All *c-roots* (a.k.a. *identities*) and the third column σ_1 component of any matrix have the reference rotator as a factor.

The DyCon3 addition rules are easier to follow by first considering an example. Refer to the second matrix of the COI opposite triplet. Its second row is the local reference row; δ_1 is the only rotator used in the row, and it is the matrix reference rotator; its c-root is related to x_1; and it is found by adding the row terms $\{\delta_1 C_1, \delta_1 C_2, \delta_1 \sigma_1\}$ directly as they are (that is, no rotations) to get its reference c-root $\delta_1 x_1$. All the matrix third-column σ_1 components have the reference rotator δ_1 as a factor. The first-row c-root is found by rotating C_1 by δ_3 and C_2 by δ_5 (C_1 and C_2 are counter rotated) and then adding the results to $\delta_1 \sigma_1$ to get its c-root $\delta_1 x_0$. The third-row c-root is found by rotating C_1 by δ_5 and C_2 by δ_3 (C_1 and C_2 are again counter rotated but in the opposite directions as previously) and then adding the results to $\delta_1 \sigma_1$ to get its c-root $\delta_1 x_2$.

Generic rules applicable to all matrices are as follows:

1. Find the row with x_1 (or $-x_1$) as its c-root and add its rotated components as they are to get the first c-root.
2. Combine the matrix's other two non-reference rotators as two ordered pairs. For the example above, they are $\{\delta_3, \delta_5\}$ and $\{\delta_5, \delta_3\}$.
3. Perform the scalar product of $\{C_1, C_2\}$ with the first pair of rotators; then add the result to the matrix rotated (by the reference rotator) third component to obtain a second c-root.
4. Perform the same steps as above using the second pair of rotators to obtain the third c-root.

The steps above are designated as the *COI upgraded matrix sextuplet addition rules*.

CHAPTER 3

TOTAL CUBIC STRUCTURE:
COI QUADRATIC WITH REAL ROOTS

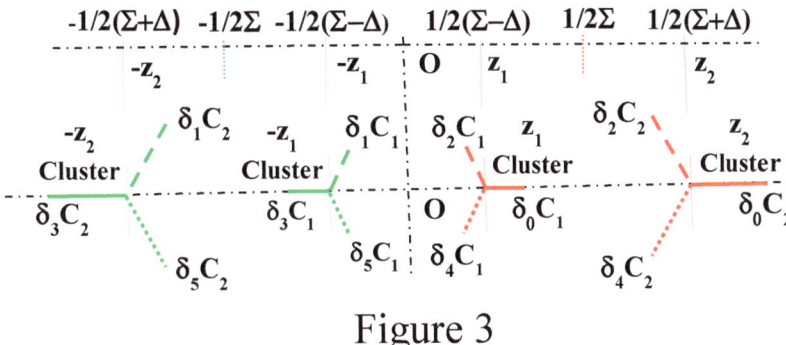

Figure 3

The q-root pairs $\{-z_2, -z_1\}$ and $\{z_2, z_1\}$ of both the COI and its opposite quadratic are depicted on the top real axis of figure 3. To avoid overlapping 3-clusters, the roots of each quadratic are assumed to be distinct and to have the same sign.

The COI coefficients are assumed to be real. Then, the COI *quadratic* b_1 coefficient is real also. Its q-root midpoint is negative if b_1 is positive and it is positive if otherwise.

Q-roots can be real or complex. The case of real roots is considered first.

Q-roots are 3-clusters' power elements that are currently assumed to be real; hence, one root element of each cluster is also real. The root elements are symmetrically located about a center, the ultimate true location of which is yet to be determined. In figure 3, the center is associated with its 3-cluster power element just as a convenience.

The first DyCon3 matrix that follows is associated with the COI, and the second is associated with the COI opposite cubic. Their quadratics differ only by the sign of their b_1 coefficient.

$$[\delta_0 C_1 + \delta_0 C_2 + \delta_0 \sigma_1] = 3\,\delta_0 x_1$$
$$[\delta_4 C_1 + \delta_2 C_2 + \delta_0 \sigma_1] = 3\,\delta_0 x_2$$
$$[\delta_2 C_1 + \delta_4 C_2 + \delta_0 \sigma_1] = 3\,\delta_0 x_0$$

$$[\delta_3 C_1 + \delta_3 C_2 + \delta_3 \sigma_1] = 3\,\delta_3 x_1$$
$$[\delta_1 C_1 + \delta_5 C_2 + \delta_3 \sigma_1] = 3\,\delta_3 x_2$$
$$[\delta_5 C_1 + \delta_1 C_2 + \delta_3 \sigma_1] = 3\,\delta_3 x_0$$

In figure 3, the real root elements are shown as the red $\delta_0 C_1$ and $\delta_0 C_2$ segments for the quadratic with a negative b_1 coefficient and as the green $\delta_3 C_1$ and $\delta_3 C_2$ segments for the other quadratic.

The other two root elements of each 3-cluster are complex conjugate pairs and are depicted as the red $\{\delta_2 C_1, \delta_4 C_1\}$ and $\{\delta_2 C_2, \delta_4 C_2\}$ for the quadratic with the negative b_1 coefficient and as the green $\{\delta_1 C_1, \delta_5 C_1\}$ and $\{\delta_1 C_2, \delta_5 C_2\}$ for the other quadratic.

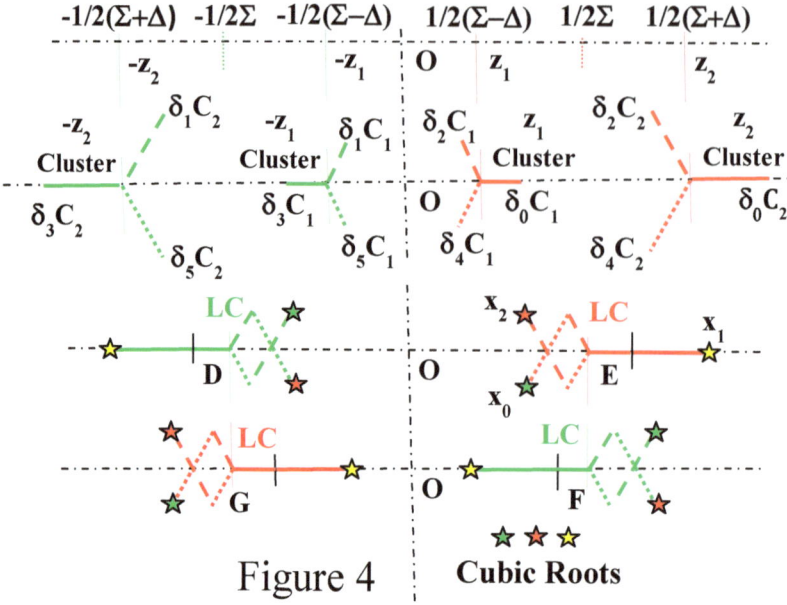

Figure 4   **Cubic Roots**

Figure 4 is an extension of figure 3, and it is bound by the same assumptions. It depicts how local *c-root patterns* for the COI and for its opposite cubic are first generated and then moved to their true locations.

A *local c-root pattern* is defined as the c-root pattern obtained by assembling root elements of the 3-clusters associated with the symmetrical columns of a DyCon3 matrix in accordance with the DyCon3 addition rules. A *global c-root pattern* is defined as the sum of an L-root and a local c-root pattern.

Red and green LC (local center) points are positioned on the bottom two axes of figure 4 to act as temporary reference points, about which clusters of root elements are assembled to create local c-root patterns. Their actual location is determined by the signs of both the *COI quadratic* b_1 coefficient and of the COI *a* coefficient. The first determines the location of ½ Σ, which is the midpoint of the *COI quadratic* roots, as shown in figure 4, top row. The second determines the actual location of the center of the COI local c-root pattern, as discussed later.

The first DyCon3 matrix that follows is associated with the COI, and the second is associated with the *COI opposite* cubic. These are used as aids to graphically build local and global c-root patterns and to determine their true locations.

$$[\delta_0 C_1 + \delta_0 C_2 + \delta_0 \sigma_1] = 3\,\delta_0 x_1 \qquad\qquad [\delta_3 C_1 + \delta_3 C_2 + \delta_3 \sigma_1] = 3\,\delta_3 x_1$$
$$[\delta_4 C_1 + \delta_2 C_2 + \delta_0 \sigma_1] = 3\,\delta_0 x_2 \qquad\qquad [\delta_1 C_1 + \delta_5 C_2 + \delta_3 \sigma_1] = 3\,\delta_3 x_2$$
$$[\delta_2 C_1 + \delta_4 C_2 + \delta_0 \sigma_1] = 3\,\delta_0 x_0 \qquad\qquad [\delta_5 C_1 + \delta_1 C_2 + \delta_3 \sigma_1] = 3\,\delta_3 x_0$$

At first sight, the COI DyCon3 matrix appears to be associated with the red items on the middle two real axes. This is not true because any and all the quantities—C_1, C_2, and σ_1—can be negative, as determined by the coefficients b_1 and a; therefore, their actual location may be the additive inverse of the location suggested by their associated rotators.

On all real axes, items associated with the negative quadratic b_1 coefficient are shown in red, and those associated with the other quadratic are shown in green.

Right-side identities (a.k.a. c-roots) are computed in accordance with the DyCon3 addition rules. The rules have already been described at length, but they are repeated here in a somewhat different form to help better visualize how local and global c-root patterns are graphically created and positioned.

A red, local c-root pattern (LCRP) is obtained in three steps as follows:

1. The center of the z_1 cluster is positioned over the red LC point of the third real axis. The real LCRP c-root for the red items is obtained by adding the real red root element $\delta_0 C_2$ to the real red root element $\delta_0 C_1$ as they are (no rotations). This c-root is shown with a yellow halo. It is associated with the c-root x_1 only as a result of choosing the first row of the original DyCon3 matrix as the global-reference row, and not from a feature that is unique to x_1. The row can be associated with the first row of either the left or the right matrices.

2. A complex LCRP c-root for the red items is obtained by adding the complex red root element $\delta_2 C_2$ to the complex red root element $\delta_4 C_1$. Note that $\delta_2 C_2$ is obtained via a CCW $2\pi/3$ rotation of $\delta_0 C_2$ and that $\delta_4 C_1$ is obtained by a CW $2\pi/3$ rotation of $\delta_0 C_1$ (that is, C_2 and C_1 are counter rotated) before being added. This c-root is shown with a red halo. It is associated with the c-root x_2 and with either the second row of the left or of the right matrices on the opposite page.

3. The other complex LCRP c-root for the red items is obtained by adding the red root element $\delta_4 C_2$ to the red root element $\delta_2 C_1$. Again, note that $\delta_4 C_2$ is obtained by a CW $2\pi/3$ rotation of $\delta_0 C_2$ and that $\delta_2 C_1$ is obtained by a CCW $2\pi/3$ rotation of $\delta_0 C_1$ before being added. (Again, C_2 and C_1 are counter rotated but in directions opposite to those used under two above). This last c-root is shown with a green halo. The root is associated with the c-root x_0 and the third row of either the left or right matrices on the opposite page.

4. The overall result is the red, local c-root pattern shown on the third real axis of figure 4.

The same red, local c-root pattern is obtained if the roles of the z_1 and z_2 clusters are reversed—that is, if the center of the z_2 cluster is positioned over the red LC point and the root elements of the z_1 cluster are added to them. Exchanging the roles of the z_1 and z_2 clusters is the graphical equivalent of interchanging symmetrical columns in said DyCon3 matrices.

The green, local c-root pattern to the left of the third axis can be generated by using the π *correspondence* between its parts and those of the red pattern. It can also be assembled via the same procedure used to assemble the red pattern. Of course, the names, colors, and signs for the power and root elements of the clusters, as well as those of the other green items, will need to be changed as appropriate.

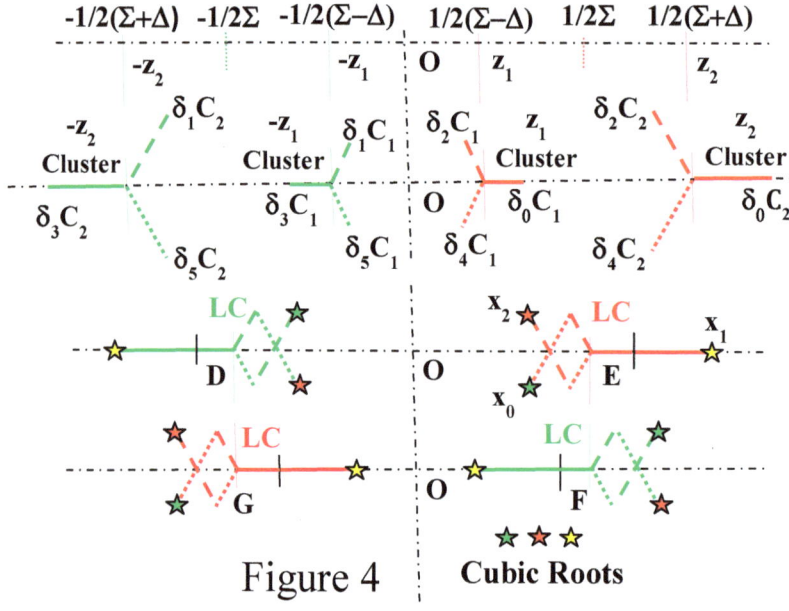

Figure 4 | Cubic Roots

$$[\delta_0C_1 + \delta_0C_2 + \delta_0\sigma_1] = 3\,\delta_0x_1$$
$$[\delta_4C_1 + \delta_2C_2 + \delta_0\sigma_1] = 3\,\delta_0x_2$$
$$[\delta_2C_1 + \delta_4C_2 + \delta_0\sigma_1] = 3\,\delta_0x_0$$

$$[\delta_3C_1 + \delta_3C_2 + \delta_3\sigma_1] = 3\,\delta_3x_1$$
$$[\delta_1C_1 + \delta_5C_2 + \delta_3\sigma_1] = 3\,\delta_3x_2$$
$$[\delta_5C_1 + \delta_1C_2 + \delta_3\sigma_1] = 3\,\delta_3x_0$$

Figure 4, the COI, and the COI opposite matrices are shown again for convenience.

If the COI quadratic b_1 coefficient is negative, then the third-axis red, local c-root pattern pertains to the COI and the green, local c-root pattern pertains to the COI opposite cubic. If b_1 is positive, then the green, local c-root pattern pertains to the COI and the red, local c-root pattern pertains to the COI opposite cubic.

Assume b_1 to be negative. Then:

If the COI a coefficient is negative, σ_1 is positive and can be represented by the third-axis OE segment. Moving the red point LC to coincide with the point E generates the COI global c-root pattern. If the COI a coefficient is positive, then σ_1 is negative and can be represented by the bottom-axis OG segment. Moving the red, local c-root pattern point LC to coincide with point G generates the COI global c-root pattern.

The value of σ_1 has been chosen large enough to avoid overlapping c-root patterns.

If the b_1 coefficient of the COI quadratic is positive, then the third-axis green pattern is the proper COI local c-root pattern. It implies the following:

If the COI a coefficient is positive, σ_1 is negative and can be represented by the third-axis OD segment. Moving the green point LC to coincide with point D generates the COI global c-root pattern. If the COI a coefficient is negative, then σ_1 is positive and can be represented by the bottom-axis OF segment. Moving the green point LC to coincide with point F generates the COI global c-root pattern.

A Complete COI Symmetrical Structure

Figure 5 depicts a complete and *fully symmetrical cubic structure*. It contains the roots of the COI, those of its *symmetrically related* cubics, and those of the three cubics derived from the COI *opposite* quadratic for a total of six cubics and eighteen roots. The roots of the COI triplet and those of the COI *opposite-matrix* triplet are bound by a π *correspondence*.

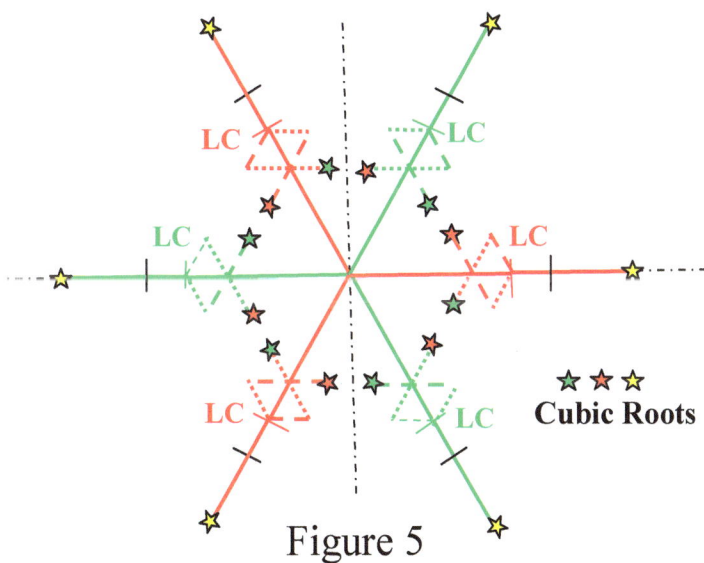

Figure 5

In figure 5, horizontal, global c-root patterns are identical to those on the third axis of figure 4 and are subject to the same assumptions. The additional patterns were generated by rotating the horizontal patterns around the figure center by 120° and 240° in the same direction (CW or CCW) as per the properties of DyCon3 matrix triplets. Rotations can be by 120° in opposite directions. Other ways to generate the figure will be described later.

The structure contains six global c-root patterns, with each linked to a cubic and to a DyCon3 matrix. Parts derived from the COI and those derived from the COI opposite cubic are treated separately because they were developed at different times and also to emphasize features associated with different quadratics. They are, and should be treated as, integral parts of one indivisible structure. In particular, each *c-root*, *global*, and *local c-root pattern* should be considered as having *correspondent* parts circularly located 60° (or multiples thereof) around the figure's center. The parts of the structure have several types of correspondence: $2\pi/3$ (within the COI and its *matrix* triplet and within the COI opposite cubic and its *matrix* triplet), π (between the COI and its *opposite* quadratic), and $\pi/3$ (by combining the other two). As a result, every part has an additive inverse, and the total sum of all the parts, referenced to the figure center, is equal to zero.

The *fully symmetrical cubic structure* shows no change if it is rotated by any multiple of $\pi/3$ radians except for the color. Sextuplets of cubics with this feature are said to form a *family of cubics*.

A figure similar to figure 5 is generated using figure 4 bottom patterns. Their only difference would be the location and the orientation of the red and green c-root patterns.

COI Symmetrical Structure: Alternate Assembly

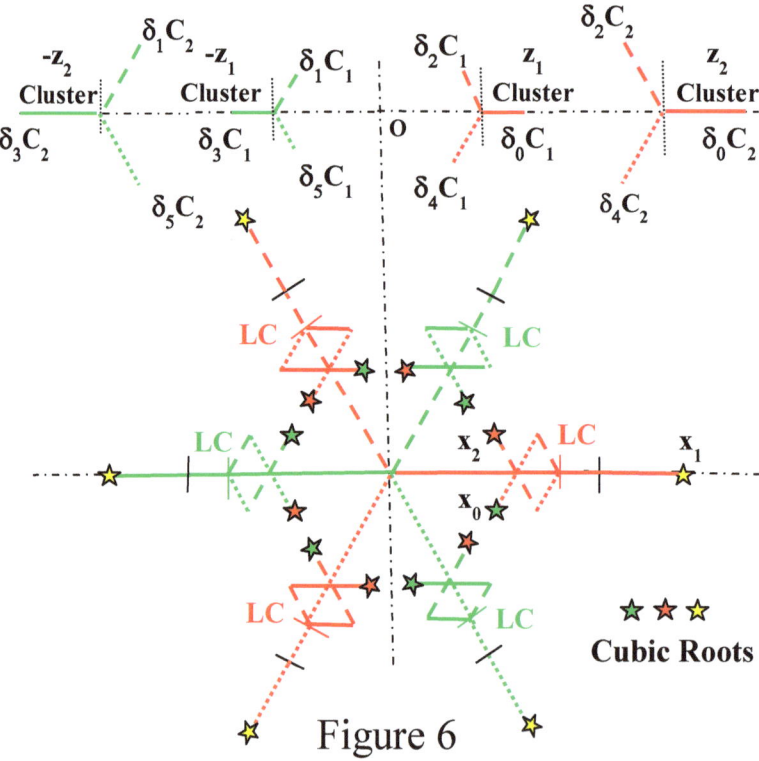

Figure 6

In figure 6 the addition rules of DyCon3 matrices are applied directly to the root elements of the second real axis of figure 4 (repeated at the top of the figure) to assemble the fully symmetrical cubic structure.

The COI upgraded matrix sextuplet is repeated as a convenience, even though it is not used except to verify that the new procedure conforms to its addition rules.

$[\delta_0 C_1 + \delta_0 C_2 + \delta_0 \sigma_1] = 3\delta_0 x_1$ $[\delta_3 C_1 + \delta_3 C_2 + \delta_3 \sigma_1] = 3\ \delta_3 x_1$

$[\delta_4 C_1 + \delta_2 C_2 + \delta_0 \sigma_1] = 3\delta_0 x_2$ $[\delta_1 C_1 + \delta_5 C_2 + \delta_3 \sigma_1] = 3\ \delta_3 x_2$

$[\delta_2 C_1 + \delta_4 C_2 + \delta_0 \sigma_1] = 3\delta_0 x_0$ $[\delta_5 C_1 + \delta_1 C_2 + \delta_3 \sigma_1] = 3\ \delta_3 x_0$

$[\delta_0 C_1 + \delta_2 C_2 + \delta_4 \sigma_1] = 3\delta_4 x_0$ $[\delta_3 C_1 + \delta_5 C_2 + \delta_1 \sigma_1] = 3\ \delta_1 x_0$

$[\delta_4 C_1 + \delta_4 C_2 + \delta_4 \sigma_1] = 3\delta_4 x_1$ $[\delta_1 C_1 + \delta_1 C_2 + \delta_1 \sigma_1] = 3\ \delta_1 x_1$

$[\delta_2 C_1 + \delta_0 C_2 + \delta_4 \sigma_1] = 3\delta_4 x_2$ $[\delta_5 C_1 + \delta_3 C_2 + \delta_1 \sigma_1] = 3\ \delta_1 x_2$

$[\delta_0 C_1 + \delta_4 C_2 + \delta_2 \sigma_1] = 3\delta_2 x_2$ $[\delta_3 C_1 + \delta_1 C_2 + \delta_5 \sigma_1] = 3\ \delta_5 x_2$

$[\delta_4 C_1 + \delta_0 C_2 + \delta_2 \sigma_1] = 3\delta_2 x_0$ $[\delta_1 C_1 + \delta_3 C_2 + \delta_5 \sigma_1] = 3\ \delta_5 x_0$

$[\delta_2 C_1 + \delta_2 C_2 + \delta_2 \sigma_1] = 3\delta_2 x_1$ $[\delta_5 C_1 + \delta_5 C_2 + \delta_5 \sigma_1] = 3\ \delta_5 x_1$

The procedure is described in detail only in reference to the first matrix of the first column of the sextuplet. It consists of the following steps:

1. Choose δ_0 as the reference rotator and draw $\delta_0 \sigma_1$ as its L-root. This defines the red horizontal LC point. Add $\delta_0 C_1$ and $\delta_0 C_2$ as they are (no rotations) to $\delta_0 \sigma_1$ and get the c-root x_1, shown with a yellow halo.

2. Combine the non-reference rotators δ_4 and δ_2 as the ordered pairs (δ_4, δ_2) and (δ_2, δ_4).

3. Perform the scalar product of the first ordered pair and the (C_1, C_2) pair (counter rotate C_1 and C_2) to get $\delta_4 C_1$ and $\delta_2 C_2$; add them to $\delta_0 \sigma_1$ to get the c-root x_2, shown with a red halo.

4. Perform the scalar product of the second ordered pair and the (C_1, C_2) pair (counter rotate C_1 and C_2, but in the direction opposite of that used in three above) to get $\delta_2 C_1$ and $\delta_4 C_2$; add the result to $\delta_0 \sigma_1$ and get the c-root x_0, shown with a green halo.

5. Perform steps one through five above for all remaining δ_1, ..., δ_5 rotators, *mutatis mutandis*, the get the other five local and global c-root patterns.

This five step graphical procedure produces the *fully symmetrical cubic structure* by applying the DyCon3 addition rules directly to the root elements of the two associated 3-clusters. It is easy to remember and needs no reference to their associated matrices or their reference rows. Note also that step five can be accomplished by using the $2\pi/3$ correspondence among the red items and the π correspondence between the red and the green items as well as by alternate procedures.

Figure 6 reflects the assumption that the COI quadratic b_1 coefficient and the COI a coefficient are both negative as discussed on pages 30 and 31 while developing the COI global patterns.

Cubic Structure: A Dynamic View

The components in figure 5 (which is the same as the bottom of figure 6) can be imparted a dynamic behavior by using a new set of rotators $\{\mu_0, \mu_1, \mu_2, \mu_3, \mu_4, \mu_5\}$ that are designated as the μ *rotators* and are defined as follows:

$$\mu_0 = \varepsilon^{\frac{j2\pi}{6}(n+0)}; \qquad \mu_1 = \varepsilon^{\frac{j2\pi}{6}(n+1)}; \qquad \mu_2 = \varepsilon^{\frac{j2\pi}{6}(n+2)}$$

$$\mu_3 = \varepsilon^{\frac{j2\pi}{6}(n+3)}; \qquad \mu_4 = \varepsilon^{\frac{j2\pi}{6}(n+4)}; \qquad \mu_5 = \varepsilon^{\frac{j2\pi}{6}(n+5)}$$

	0		1		2		3		4		5	
0	μ_0,	$0°$	μ_1,	$60°$	μ_2,	$120°$	μ_3,	$180°$	μ_4,	$240°$	μ_5,	$300°$
1	μ_1,	$60°$	μ_2,	$120°$	μ_3,	$180°$	μ_4,	$240°$	μ_5,	$300°$	μ_0,	$0°$
2	μ_2,	$120°$	μ_3,	$180°$	μ_4,	$240°$	μ_5,	$300°$	μ_0,	$0°$	μ_1,	$60°$
3	μ_3,	$180°$	μ_4,	$240°$	μ_5,	$300°$	μ_0,	$0°$	μ_1,	$60°$	μ_2,	$120°$
4	μ_4,	$240°$	μ_5,	$300°$	μ_0,	$0°$	μ_1,	$60°$	μ_2,	$120°$	μ_3,	$180°$
5	μ_5,	$300°$	μ_0,	$0°$	μ_1,	$60°$	μ_2,	$120°$	μ_3,	$180°$	μ_4,	$240°$

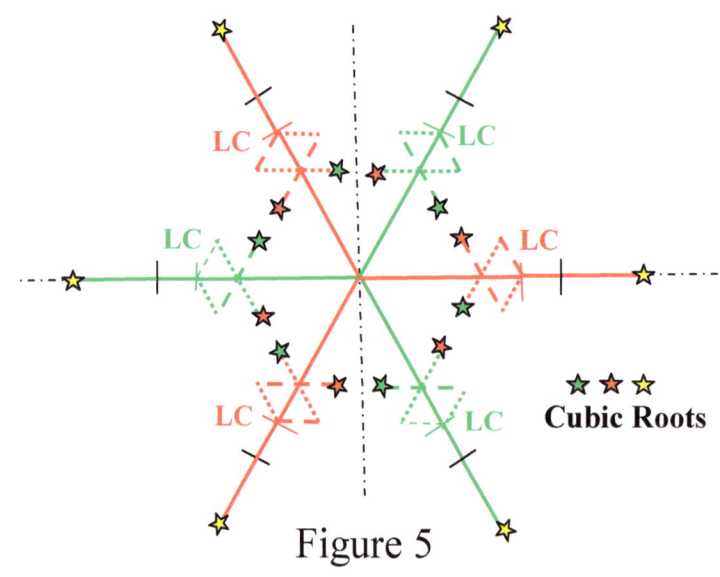

Figure 5

The opposite page table is defined as the μ *rotators' table*. Rotator subscripts are modulo six sums of the parameter n (first column) and a constant (top row). The angle of a rotator is the product of its subscript times 60°. The table cells display a rotator and the degrees of its angle, which is also the angle of the associated c-root patterns.

A μ rotator has the same angle and performs the same function as a δ rotator with equal subscript. Replacing δ rotators with μ rotators and incrementing (decrementing) n values by one imparts a dynamic behavior to the local and global patterns of figure 5.

As n increments from zero to one, every μ *rotator* morphs into the rotator circularly located at its right and the associated c-root patterns move CCW by 60°. That is,

1. μ_0 morphs into μ_1 and its associated patterns move from 0° to 60°.
2. μ_1 morphs into μ_2 and its associated patterns move from 60° to 120°.
3. μ_2 morphs into μ_3 and its associated patterns move from 120° to 180°.
4. μ_3 morphs into μ_4 and its associated patterns move from 180° to 240°.
5. μ_4 morphs into μ_5 and its associated patterns move from 240° to 300°.
6. μ_5 morphs into μ_0 and its associated patterns move from 300° to 0°.

In other words, figure 5 red, horizontal patterns (global and local) move CCW by 60° and onto the location just vacated by green patterns, which move onto the red patterns at 120°, which move onto the green patterns at 180°, which move onto the red patterns at 240°, which move onto the green patterns at 300°, which move onto the red patterns at 0°. Figure 5 has rigidly rotated CCW by 60°. Disregarding the color coding (which is no longer relevant), the figure still looks the same. It is designated as figure 5A.

Similar results are obtained using other rotation sequences, for instance, by rotating the elements of all patterns CW by 60°. In this case, the horizontal, red patterns rotate CW by 60°— the L-root around the figure center and, at the same time, the components of the local c-root pattern around their local center (red LC point). All other patterns experience similar rotations. This case corresponds to n decrementing from zero to minus one.

As n increments from one to two, the original red, horizontal patterns now located at 60° move CCW onto the 120° location just vacated by green patterns, which move on to the 180° location just vacated by red patterns, which move onto the 240° location just vacated by green patterns, which move onto the 300° location just vacated by red patterns, which move onto the 0° location just vacated by green patterns, which move onto the 60° location just vacated by red patterns. Figure 5A has rigidly rotated CCW by 60° to become figure 5B. Disregarding the color coding, figure 5B looks like figure 5A, which looks the same as figure 5.

The combination of both rigid rotations defines a *two-step symmetrical rotation*, which is more conveniently considered as two consecutive CCW *one-step symmetrical rotations*. The first step changes figure 5 into figure 5A, and the second changes figure 5A into figure 5B. These figures are not shown; all of them would look exactly the same as figure 5, except for color.

Decrementing *n* values would cause rotations in the CW direction.

A continuous sequence of one-step symmetrical rotations moves c-roots into all their correspondent locations and it removes the apparent one-to-one relationship between a c-root and its location within the symmetrical structure.

Alternate Fully Symmetrical Structures

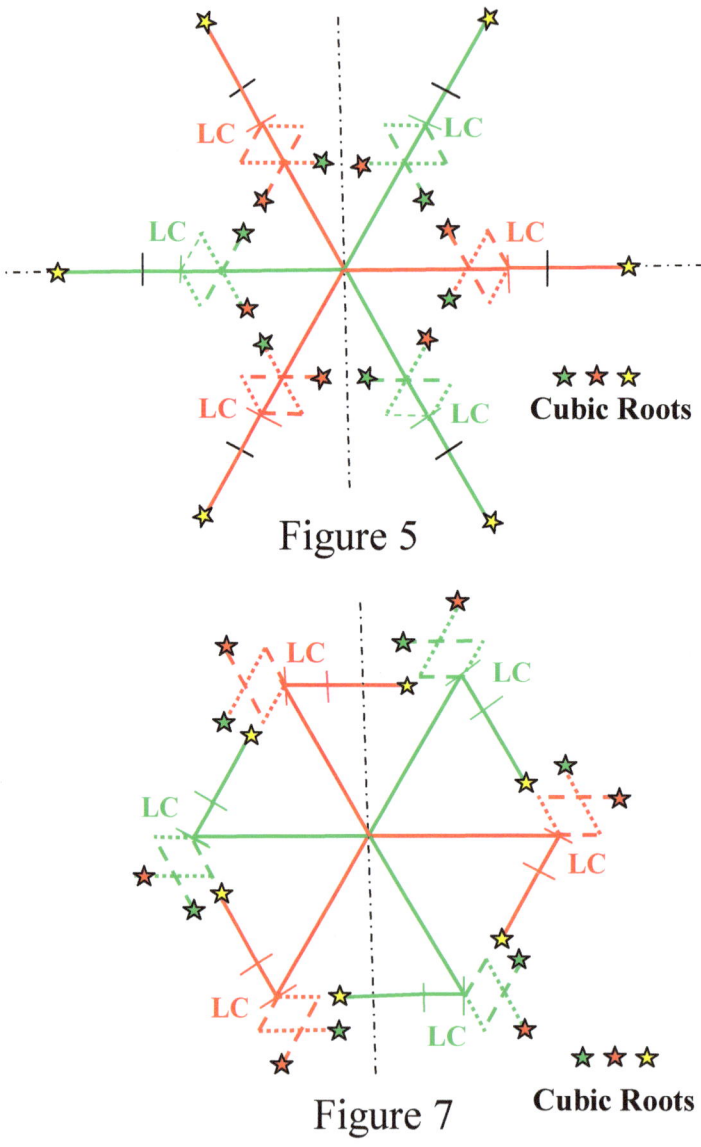

Figure 5

Figure 7

Cubic Roots

The fully symmetrical formulas allow associating any L-root with any local c-root pattern. Figure 7 depicts one such instance. The horizontal, red, local c-root pattern of figure 5 is now associated with the red L-root located at 120° CCW. The association of the other two red, local c-root patterns with L-roots has changed in a way that maintains $2\pi/3$ correspondence. Green c-root patterns have changed following the same procedure used for the red patterns (and, in a way that maintains the $2\pi/3$ correspondence).

Structures in which the COI c-root local pattern is not associated with the COI L-root are designated as non-COI structures. Their global c-root patterns look quite different from those of figures 5, which was designed to emphasize how the COI a coefficient and the COI quadratic b_1 coefficients affect the local c-root patterns of the COI family of cubics. Non-COI structures do not include the roots of the original COI.

A $120°$ CCW, in situ, rotation of local c-root patterns around their LC points generates a figure similar to figure 7 that will be referred to as figure 7-CCW. The figure is not drawn to avoid page overcrowding. The fully symmetrical structures of figure 7 would show no change if rotated by any multiple of $\pi/3$ radians except for the color. The same would be true for figure 7-CCW if drawn. That is, each depicts a *family of cubics*.

Together, they will be referred as the *COI-related family triplet*.

Figures 5, 7, and 7CCW depict the totality of cubic families that can be generated by the fully symmetric cubic formula if correctly interpreted. The formula was obtained by combining the COI cubic formula and the COI opposite-cubic formulas. All formulas are repeated below for convenience.

$$6y = -2a\sqrt[3]{1} + \sqrt[3]{4(-2a^3 + 9ab - 27c) + \sqrt{-432(a^2b^2 - 4a^3c - 4b^3 + 18abc - 27c^2)}}$$
$$+ \sqrt[3]{4(-2a^3 + 9ab - 27c) - \sqrt{-432(a^2b^2 - 4a^3c - 4b^3 + 18abc - 27c^2)}}$$

$$6y = -2a\sqrt[3]{1} + \sqrt[3]{-4(-2a^3 + 9ab - 27c) + \sqrt{-432(a^2b^2 - 4a^3c - 4b^3 + 18abc - 27c^2)}}$$
$$+ \sqrt[3]{-4(-2a^3 + 9ab - 27c) - \sqrt{-432(a^2b^2 - 4a^3c - 4b^3 + 18abc - 27c^2)}}$$

$$6y = -2a\sqrt[3]{1} + \sqrt[3]{4(-2a^3 + 9ab - 27c)\sqrt{1} + \sqrt{-432(a^2b^2 - 4a^3c - 4b^3 + 18abc - 27c^2)}}$$
$$+ \sqrt[3]{4(-2a^3 + 9ab - 27c)\sqrt{1} - \sqrt{-432(a^2b^2 - 4a^3c - 4b^3 + 18abc - 27c^2)}}$$

The formulas above refer respectively to the COI cubic formula, the COI opposite-cubic formula, and to the combined formula.

The correct interpretation of the combined formula requires that the value of the square root of unity (+1 or −1) must be held constant for all the cubics in the COI triplet and for all the cubics in the opposite-matrix triplet, but it must be different for each triplet. This limitation was designated as the *COI opposite quadratics' restriction* (page 19).

The COI cubic formula by itself produces twenty-seven roots,—three roots for each cubic radical. The COI opposite-cubic formula by itself also produces twenty-seven roots for a total of fifty-four roots.

The combined formula if correctly interpreted also produces fifty-four roots.

Figure 5, figure 7 and figure 7-CCW include 18 roots each for a total of fifty-four roots. Therefore, the total number of roots generated by the formulas and those depicted by the fully symmetrical structures in figures 5, 7 and 7-CCW are equal, as it should be.

In figures 5, 7 and 7-CCW for every cubic root generated by the COI formula, there is a correspondent generated by the COI opposite-cubic formula that is symmetrical to it with respect to the figure's center. Every c-root, therefore, has a diametrically opposed c-root with which it forms an additive inverse, making their sum equal to zero. All three fully symmetrical structures preserve the zero-sum property with respect to their center.

C-roots of opposite pairs do not vanish: their arrangement is such that their sum acts like the center of mass of a distribution of mass in space. This center then behaves as if all the roots were located on it and formed some kind of local black hole.

It is not necessary for a cubic to have real coefficients for it to generate a fully symmetrical formula that contains one or more real c-roots. For instance, consider the red, global c-roots pattern located at −120° in figure 5. The value of its L-root is obviously a complex number, which means that its cubic has at least one complex coefficient. If that cubic is used as a new COI, it generates the same fully symmetrical cubic structures and the same fully symmetrical formulas as does the original COI; therefore, the roots of some cubics with complex coefficients can be parts of a fully symmetrical cubic structure that includes some real roots.

The next chapter deals with the cases in which the roots of the COI quadratic are not real but are complex conjugate.

CHAPTER 4

TOTAL CUBIC STRUCTURE:
COI QUADRATIC WITH COMPLEX ROOTS

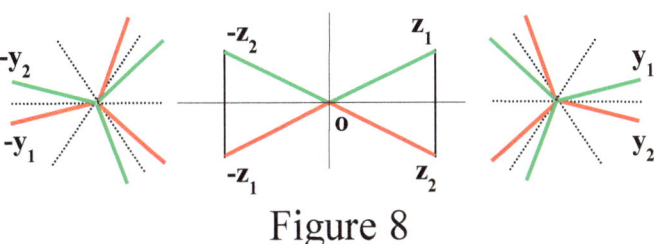

Figure 8

Two pairs of complex conjugate roots are depicted at the center of figure 8, one each for the COI and for its opposite quadratic. The real parts of each pair are also their root midpoints. Midpoints of different pairs are additive inverses. Imaginary parts for each pair are symmetrical about their midpoints and perpendicular to the real parts.

Quadratic roots are power elements of 3-clusters. They are generically labeled by the letter z with added signs and subscripts to identify them individually. Root elements form triads that are shown on the same side of and with the same colors as their power elements. One root element of each triad is generically labeled y with signs and subscripts to match those of their respective power element.

Let m and n be sequences of incrementing (or decrementing) integers and M^3 and 3α be the magnitude and angle of the z_1 root. The power and root elements of the various 3-clusters can be written, respectively, as

$$z_1 = M^3 \varepsilon^{j(2\pi m + 3\alpha)}; \quad z_2 = M^3 \varepsilon^{j(2\pi m - 3\alpha)}$$

$$y_1 = M\varepsilon^{j\left(\frac{2\pi n}{3} + \alpha\right)}; \quad y_2 = M\varepsilon^{j\left(\frac{2\pi n}{3} - \alpha\right)}$$

$$-z_1 = M^3 \varepsilon^{j[2\pi m + 3(\pi + \alpha)]}; \quad -z_2 = M^3 \varepsilon^{j[2\pi m + 3(\pi - \alpha)]}$$

$$-y_1 = M\varepsilon^{j\left[\frac{2\pi n}{3} + (\pi + \alpha)\right]}; \quad -y_2 = M\varepsilon^{j\left[\frac{2\pi n}{3} + (\pi - \alpha)\right]}$$

The preceding relationships show that power-element pairs are symmetrical about the real axis; labeled root-element pairs inherit this symmetry and replicate it for both pairs of power elements at angular distances equal to multiples of $2\pi/3$ around a common center.

The values of the m and n parameters must be synchronized and n must assume three times as many values as m to maintain the proper correspondence among the power and root elements of each 3-cluster.

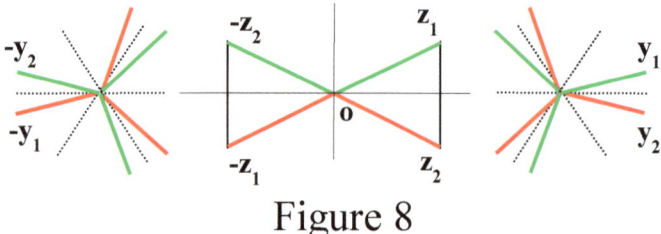

Figure 8

Figure 8 is repeated for convenience.

Located on the right side of the figure are two 3-clusters with power elements $\{z_1, z_2\}$. Associated root elements labeled $\{y_1, y_2\}$ are shown as being symmetrical about the real axis. They correspond to a modulo-three, zero value for the n parameter in the relations on the preceding page. The other two root elements of each cluster are determined by the $2\pi/3$ correspondence among 3-cluster root elements. They are symmetrical about straight lines passing through the common cluster center and are located at 120° and 240°.

A similar but opposite pattern is shown on the left side of figure 8 to depict the other two 3-clusters with power elements $\{-z_1, -z_2\}$. One pair of root elements is labeled $\{-y_1, -y_2\}$ and is shown as being symmetrical about the real axis.

The labeled root element pairs are used to generate *local c-root patterns* for the COI and/or its opposite cubic as shown in figure 9. They are based on the generic DyCon3 addition rules that were described in detail regarding figure 4 (page30) and figure 6 (page 32), both of which consider the case of quadratics with real roots.

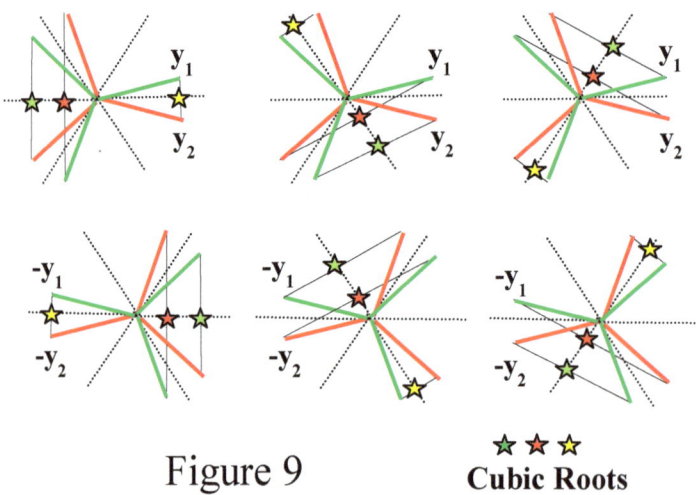

Figure 9

★ ★ ★
Cubic Roots

The top row of Figure 9 depicts the local c-root patterns for figure 8 right side triads; the bottom row depicts those for the left side. If the COI quadratic b_1 coefficient is negative, then the COI q-root midpoint is positive, the right side of figure 8 and the top row of figure 9 all relate to COI items. C-roots are shown at one half their true distances from their 3-cluster centers.

Other ways to generate local c-root patterns are similar to those already described with regard to quadratic with real roots.

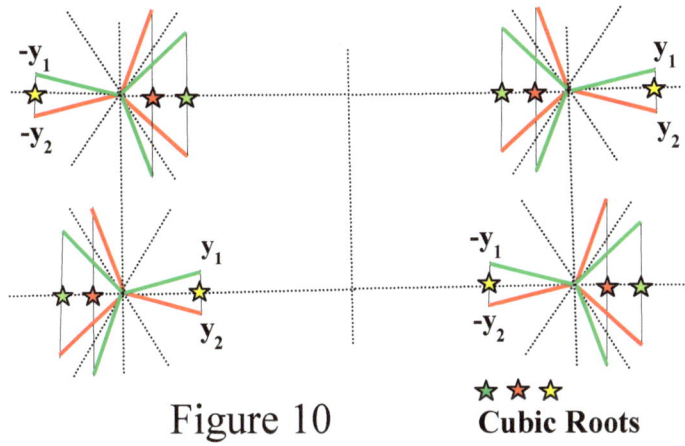

Figure 10   ★ ★ ★
                            Cubic Roots

Figure 10 shows the effect that the COI a coefficient and the COI quadratic b_1 coefficient have on the local and the global c-root patterns of the COI family real roots.

To form a global c-root pattern, the centers of the root elements' triads are located on the real axis at a distance from the imaginary axis that is equal to the magnitude of a real L-root, which has been chosen large enough to create graphics with easy-to-visualize properties.

If the COI quadratic b_1 coefficient is negative, the correct COI local c-root pattern is the first one on the top row of figure 9. If the COI a coefficient is negative too, then the COI real L-root is positive, and the top-right global c-root pattern of figure 10 correctly depicts the COI cubic roots; however, if a is positive, then the COI real L-root is negative, and if added to the local c-root pattern, it moves it to the left, causing the bottom-left global root pattern of figure 10 to correctly depict the COI cubic roots.

If b_1 is positive, then the correct COI local c-root pattern is the first one on figure 9 bottom row. If a is positive too, then the COI real L-root is negative, and the top-left global root pattern of figure 10 correctly depicts the COI cubic roots; however, if a is negative, then the COI real L-root is positive, and if added to the local c-root pattern, it moves it to the right, causing the bottom-right, global root pattern of figure 10 to correctly depict the COI cubic roots.

In all cases, the correct root patterns for the COI opposite cubic are the patterns that are symmetrical to the correct COI root patterns with respect to the imaginary axis. This feature causes the roots of their global patterns to form additive inverses and the sum of all their c-roots with respect to the figure center to be equal to zero.

The top patterns of figure 10 are chosen next for creating a fully symmetrical cubic structure for a COI with three positive real roots as shown by figure 11 below.

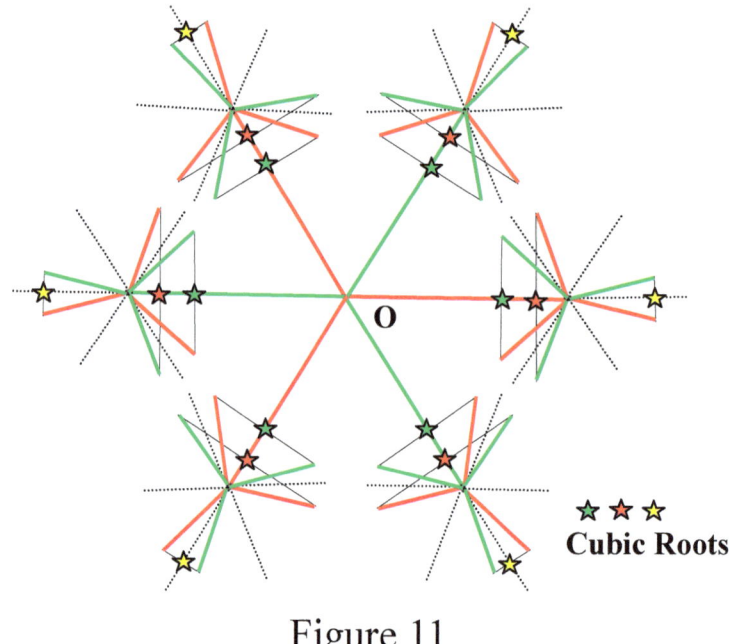

Cubic Roots

Figure 11

The figure starts with the top patterns of figure 10 and then rotates them by $2\pi/3$ in both directions. The figure has all features of and a structure similar to that of figure 5; therefore, it is a fully symmetrical cubic structure. It is considered to be the combination of two triplets of cubics, the parts of which have a $2\pi/3$ correspondence within a triplet and a π correspondence with the parts of different triplets.

A similar figure can be assembled using the bottom patterns of figure 10.

Alternate Parts Arrangements within Formulas

Figure 11 is not unique because, within a triplet, any of its three L-roots can be associated with any of its local c-root patterns. Only three fully symmetrical cubic structures can be obtained because once an L-root and a local c-root pattern have been chosen, the other two within the same triplet and the other three within the other triplet are determined by the correspondence of parts within the fully symmetrical cubic structure. The resulting three fully symmetrical cubic structures look different, and two of them do not contain the COI roots.

The other two such associations are shown as figures 12 and 13. They are obtained from figure 11 by rotating *in situ* all local cubic c-root patterns by $2\pi/3$, either in the CW (figure 12) or in the CCW (figure 13) directions. Other ways to obtain them are similar to those used for real q-roots in analogous cases.

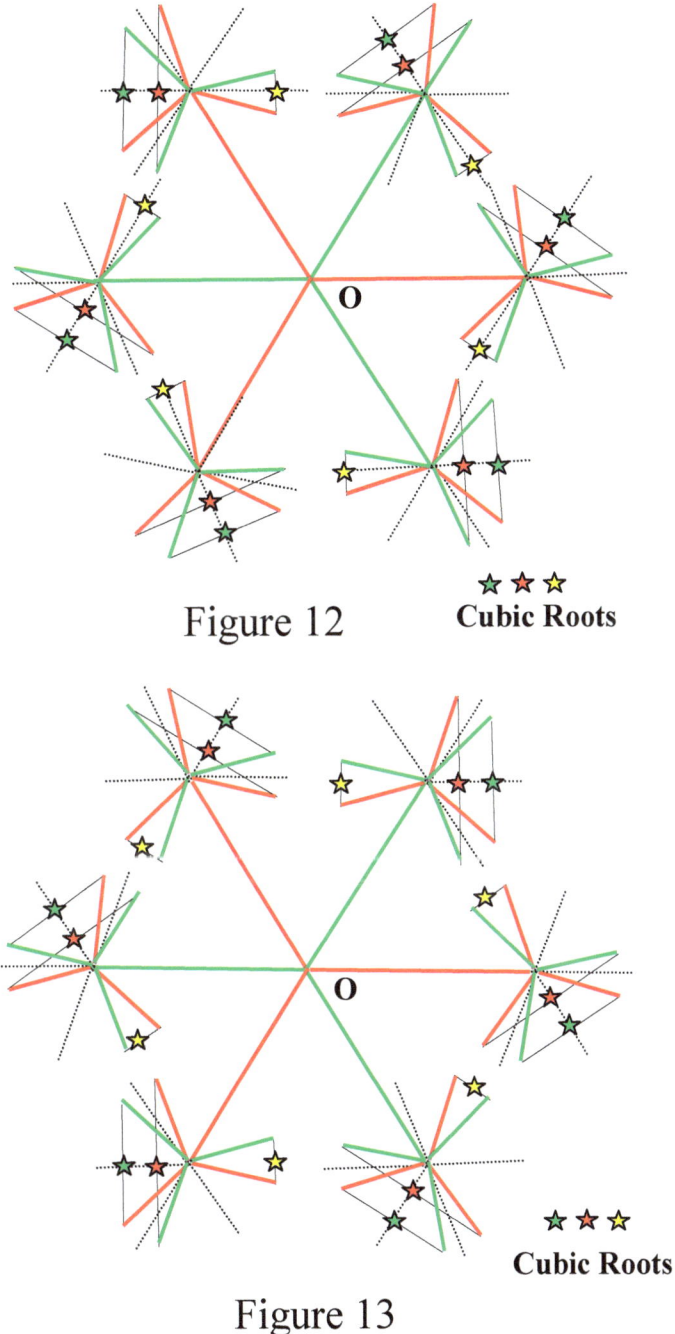

Figure 12

★ ★ ★
Cubic Roots

Figure 13

★ ★ ★
Cubic Roots

Figures 12 and 13 do not contain the COI roots. Both feature global c-root pattern pairs circularly located 180° apart around the figure center. An important consequence of this is that for every cubic root in the structure, there is another one that is diametrically opposite to it; the sum of all the roots, with respect to the structure center, is then zero. This property was designated as the *center zero sum property*.

Other Complex Roots of Opposite Quadratics

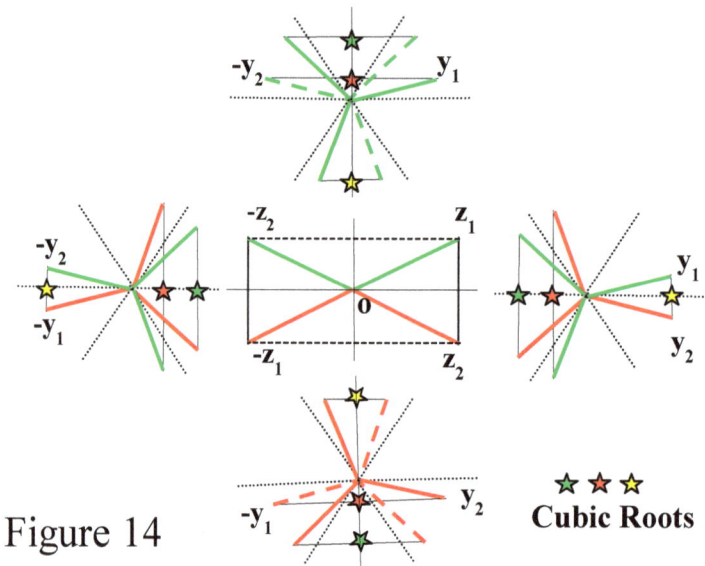

Figure 14

Cubic Roots

Figure 14 is an expanded version of figure 8. Depicted at the center are the complex conjugate roots of the *COI quadratic* and of its *opposite quadratic*. At the sides are depicted local root patterns associated with either the COI or its *opposite* cubic. At the top and at the bottom are depicted two new local root patterns.

The four q-roots $\{-z_2, -z_1, z_2, z_1\}$ can be grouped as root pairs that are symmetrical to the real axis [$\{-z_2, -z_1\}$ and $\{z_1, z_2\}$] or to the imaginary axis [$\{-z_2, z_1\}$ and $\{-z_1, z_2\}$] or to the figure center [$\{-z_2, z_2\}$ and $\{-z_1, z_1\}$]. They are generically designated as *z-pairs*.

The first group of z-pairs has already been considered. It is characterized by root pairs, the real parts of which are equal to one-half of their sums (their midpoints), and the imaginary parts are equal to one-half of their differences and are additive inverses. The real parts of the different pairs are also additive inverses.

The second group of z-pairs is characterized by root pairs, the imaginary parts of which are equal to one-half of their sums (their midpoints). Their real parts are equal to one-half of their differences and are additive inverses. The imaginary parts of the different pairs also are additive inverses. The roles that the real and imaginary parts play in the second group are reversed from those that they play in the first group. Roots in this pair cannot be roots of quadratics with real coefficients. It is being considered solely for its interesting features and a possible case if the values of the cubic equation coefficients are not restricted to real numbers.

The third group is characterized by *z-root* pairs in which both real and imaginary parts are additive inverses. Their characteristics will be described later.

The roots of any z-pair in a group can be treated as roots of a quadratic that can be solved, starting with their own DyCon2 matrices. The result will be a pair of *symmetrical quadratic formulas* and an entire symmetrical quadratic structure that include the roots of the chosen z-pair and, because of symmetry, also the roots of the associated z-pair.

In figure 14, the power elements of the first z-pair of the second group are shown above the real axis (in green); their root elements are shown at the top with dashed lines for the left side and solid lines for the right side. The power elements of the second z-pair are shown below the real axis (in red); their root elements are shown at the bottom with dashed lines for the left side and solid lines for the right side.

Local c-root patterns for the second group can be found by adding root elements that are symmetrical about the imaginary axis and have different line styles. C-roots are shown at one-half of their true distance from the 3-cluster centers.

Except for the 90° rotation, the local c-root patterns of both the first and the second group of pairs have very similar shapes, as shown in figures 15 (below) and 11 (page 42). If the real and imaginary parts of the quadratic roots are equal, then the local c-root patterns would also be exactly the same except for a 90° rotation.

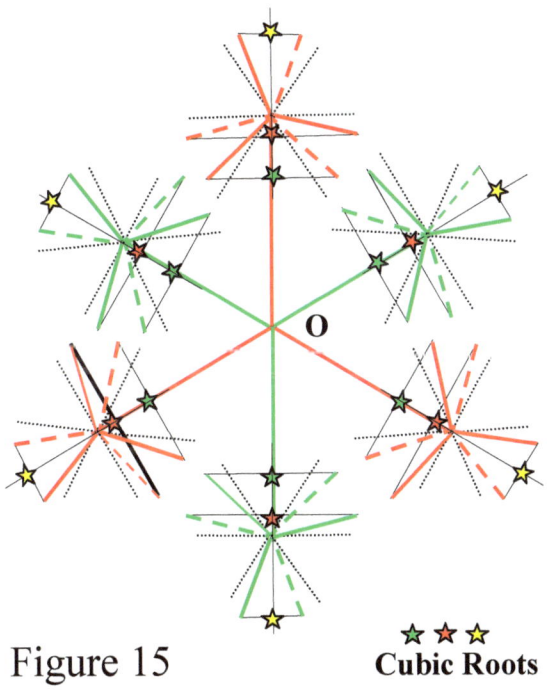

Figure 15   ★ ★ ★
                            Cubic Roots

Figure 15 shows a fully symmetrical cubic structure for the second group of pairs. The structure does not and cannot include the COI c-roots. A similar figure is obtained by interchanging diametrically opposed local patterns.

The figure is not unique because, within each triplet, any of its three L-roots can be associated with any of its local c-root patterns. Again, the total number of fully symmetrical cubic structures that can be obtained is three because once an L-root and a local root pattern are chosen, the other two within the same triplet and the other three within the other triplet are uniquely determined by the correspondences of parts within the fully symmetrical cubic structure.

Quadratic with Opposite Complex Roots

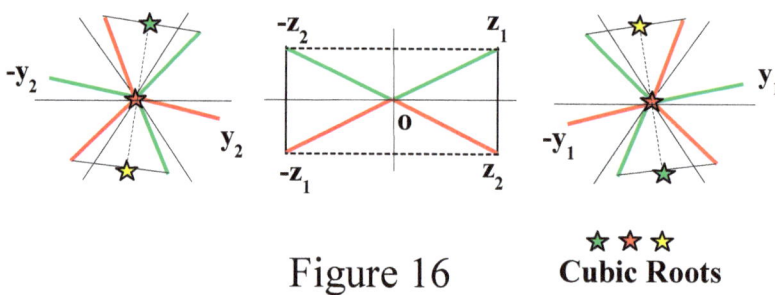

Figure 16

★ ★ ★
Cubic Roots

Figure 16 illustrates the case in which the complex roots of the COI quadratic and those of its opposite quadratic are grouped as the $\{-z_2, z_2\}$ and $\{-z_1, z_1\}$ pairs.

At the figure center are depicted the quadratic roots, which are the power elements of 3-clusters. Those in the same pair are opposite to each other.

The 3-cluster root elements associated with the $\{-z_2, z_2\}$ pair are shown on the left side of the figure; those associated with the $\{-z_1, z_1\}$ pair are shown on the right side. They are derived from the same expressions used in the first section of this chapter.

The $\{-z_2, z_2\}$ power element pair generates a first $\{-y_2, y_2\}$ root element pair; the other $\{-z_1, z_1\}$ power element pair generates a first $\{-y_1, y_1\}$ root element pair.

Adding the left side $\{-y_2, y_2\}$ root element pair as they are (no rotations) generates a first c-root that is shown with a red halo and located at the center of their 3-clusters. Rotating $-y_2$ by 120° CCW and y_2 by 120° CW and then adding the results generates a second c-root, which is shown with a yellow halo. Rotating $-y_2$ by 120° CW and y_2 by 120° CCW and then adding the results produces a third c-root, which is shown with a green halo. This 3-step procedure produces the local c-root pattern shown at the figure left side. C-roots are shown at one half of their distance from their 3-clusters center.

A similar procedure applied to the right side $\{-y_1, y_1\}$ root element pair produces the local c-root pattern shown on the figure right side.

Notice that the procedure used above to get the c-root local patterns is the same as that used for figure 6 (pages 32 and 33) and they conform to the Dycon3 addition rules.

The geometric representation shows that the cubic roots are collinear, just like those of the other two groups of z-pairs derived from quadratics with complex conjugate roots.

The location and amplitude of the cubic L-roots are not affected by any operations performed to obtain the c-root local patterns. They form a symmetrical set of six polar vectors bound by a $\pi/6$ correspondence.

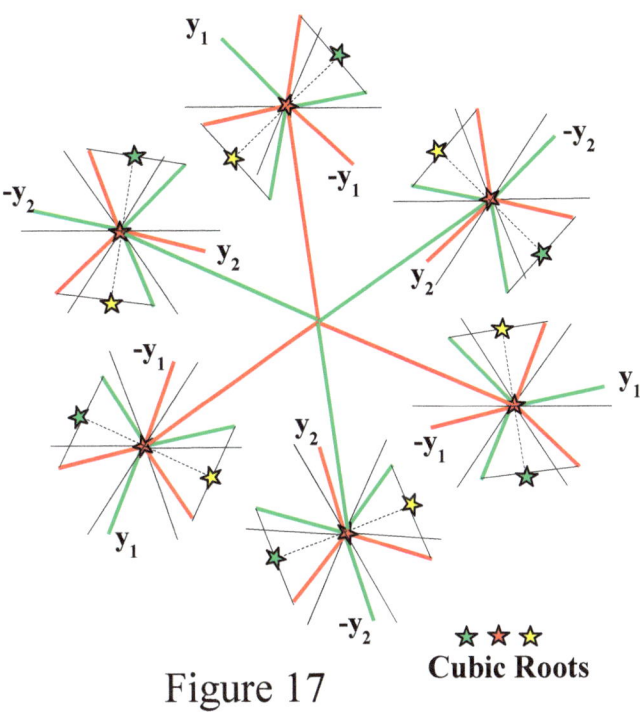

Figure 17

Cubic Roots

Figure 17 shows a possible configuration of L-roots at its center and one of three possible configurations of its global root patterns. The figure shows that every c-root has an additive inverse diametrically opposite to it. Therefore, it retains the property that the sum of all its c-roots with respect to its center is equal to zero.

As for all possible fully symmetrical cubic structures, the arrangement of parts in figure 17 is not unique, but it is one of three possible arrangements that can be obtained by rotating, in situ, CCW or CW by $2\pi/3$ all local c-root patterns. The reasons why have been explained several times already.

In general, cubic roots derived from diagonal pairs do not lie over either the real or the imaginary axis.

Intrinsic Features of a Structure

The parts of all fully symmetrical structures have relative intrinsic geometric features (such as symmetries) that are not affected by rigid motions or by changes of the coordinate system. The latter is needed only to describe the structure using algebraic methods. This action invariably adds some extraneous features and properties (such as being real or imaginary) to some specific parts. These extraneous features can appear intrinsic to the affected parts and can lead to erroneous conclusions regarding their role in the total structure. Such problems can be avoided by describing the structure independently of any coordinate system and using only its intrinsic geometric features.

This method is applied to figure 17, which is repeated below for convenience.

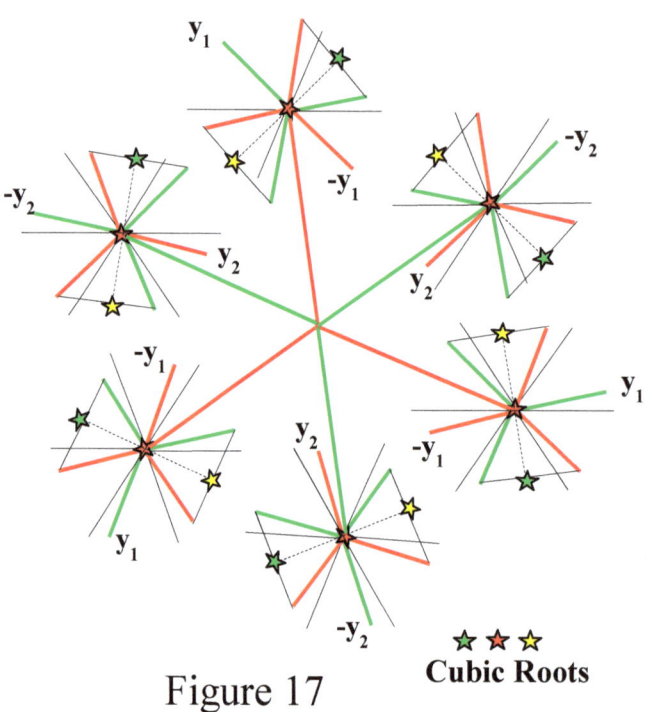

Figure 17   **Cubic Roots**

The structure has a center about which all its parts are symmetrical. It consists of six major parts designated as global c-root patterns that contain three collinear cubic roots and are radially located 60° apart around the structure center. Pairs of such patterns are diagonally opposed, which makes the sum of the entire structure with respect to the figure center equal to zero.

The whole structure contains substructures that are symmetrical about the structure center and/or about local centers. Substructures are of different types, but for each type their number is congruent with the basic structure symmetry (in this case, a $2\pi/6$ correspondence). Any substructure of one type can be joined with any substructure of another type and form a basic building block that can be replicated according to the structure-symmetry rules. All choices are admissible and are required by the definition of symmetry. Some choices may generate identical symmetrical structures; others may generate radically different symmetrical structures and may not contain the COI roots among their parts. This feature is designated as *polymorphism*.

In the present case, the substructures were identified as the L-roots and the local symmetrical patterns, which were joined together to assemble the global symmetrical patterns that were used as building blocks.

With the preceding description, no part of any kind plays any major or any different role among parts of the same kind.

The figure is characterized by its two-dimensional structure: it is then just natural that the location of most of its parts requires two-dimensional expressions. Real numbers provide only one dimension.

Formulas, Solutions and Roots

The initial symbolic formula for the *COI cubic* is rewritten below.

$$6y = -2a + \sqrt[3]{\frac{(c_1^3 + c_2^3)}{2} + \sqrt{\frac{(c_1^3 - c_2^3)^2}{4}}} + \sqrt[3]{\frac{(c_1^3 + c_2^3)}{2} - \sqrt{\frac{(c_1^3 - c_2^3)^2}{4}}}$$

The formula originally was meant to solve only the *COI cubic*. It has two flaws: (1) it produces nine solutions, whereas the COI cubic has only three roots, and (2) it is not symmetrical. Delving into the first flaw revealed that three cubics with roots bound by a $2\pi/3$ correspondence share the formula cubic radicals but do not share the other term. The formula was upgraded by changing the non-shared term from $-2a$ to

$$-2a\sqrt[3]{1}.$$

The new formula was designated as the *COI upgraded formula*. This introduced a new flaw: the new formula produces twenty-seven solutions, enough for nine cubics.

It was noticed that the radicands of the original-formula radicals are formatted as the roots produced by a nonsymmetrical quadratic formula, which was designated as the *COI quadratic*. The search for a symmetrical quadratic formula led to the COI opposite quadratic, a COI opposite cubic, and a COI opposite-matrix triplet. The initial symbolic formula for COI opposite cubic was written as

$$6y = +2a\sqrt[3]{1} + \sqrt[3]{-\frac{(C_1^3 + C_2^3)}{2} + \sqrt{\frac{(C_1^3 - C_2^3)^2}{4}}} + \sqrt[3]{-\frac{(C_1^3 + C_2^3)}{2} - \sqrt{\left(\frac{C_1^3 - C_2^3}{4}\right)^2}}$$

The preceding formula was designated as the *COI opposite formula*.

At first, the formulas were merged into one, which was designated as the *COI fully symmetrical formula*, because all its parts do have correspondent ones. In terms of the COI coefficients, it was written as

$$6y = -2a\sqrt[6]{1} + \sqrt[3]{4(-2a^3 + 9ab - 27c)\sqrt{1} + \sqrt{-432(a^2b^2 - 4a^3c - 4b^3 + 18abc - 27c^2)}}$$
$$+ \sqrt[3]{4(-2a^3 + 9ab - 27c)\sqrt{1} - \sqrt{-432(a^2b^2 - 4a^3c - 4b^3 + 18abc - 27c^2)}}$$

The symbolism used is ambiguous and confusing and needs clarifying.

The square root operation produces two results. In the quadratic formula, the symbol ± is used in front of the square root symbol as a flag to ensure that both alternatives are considered. The cubic formula uses the results of a square root operation concurrently but in different radicals. As written, each square root is meant as an absolute value, and signs are added to specify which root. Changing the signs causes the cubic radicals to change into each other.

The red square root of unity inside the cubic radicals is meant as a different type of flag. To satisfy the *COI opposite quadratics' restriction* (page 19), it must keep the same sign inside both the COI and COI opposite-matrix triplets, but it must have a different sign for different triplets. The formula does not clearly spell out this requirement.

Another problem is that each radical produces six roots, for a total of six times six times six, or 216 combinations—four times the fifty-four roots that can be produced by adding the twenty-seven c-roots from the *COI upgraded formula* to the twenty-seven c-roots from the *COI opposite formula*.

An analysis of the formula root-generation process revealed that three-fourths of the combinations should be discarded. Half of them are obtained when the square roots of unity inside different cubic radicals have the same signs. In this case, they cannot be roots of either the COI upgraded formula or the COI opposite formula. Another quarter of them are obtained when the signs of said square roots of unity are interchanged. In this case, the last two cubic radicals exchange positions, and they generate the same cubic roots, which would be counted twice.

The correct way to implement the COI fully symmetrical formula is to use both the COI upgraded formula and the COI opposite formula as a set, as follows:

$$6y = -2a\sqrt[3]{1} + \sqrt[3]{\frac{(C_1^3 + C_2^3)}{2} + \sqrt{\left(\frac{C_1^3 - C_2^3}{4}\right)^2}} + \sqrt[3]{\frac{(C_1^3 + C_2^3)}{2} - \sqrt{\left(\frac{C_1^3 - C_2^3}{4}\right)^2}}$$

$$6y = +2a\sqrt[3]{1} + \sqrt[3]{-\frac{(C_1^3 + C_2^3)}{2} + \sqrt{\frac{(C_1^3 - C_2^3)^2}{4}}} + \sqrt[3]{-\left(\frac{C_1^3 + C_2^3}{2}\right) - \sqrt{\left(\frac{C_1^3 - C_2^3}{4}\right)^2}}$$

In figure 5 and 7 (repeated on the opposite page), *local* and *global c-root patterns* derived from different formulas are shown in different colors.

For each formula, the last two radicals combine to produce three local c-root patterns bound by a $2\pi/3$ correspondence; the first radical produces three L-roots, which are also bound by a $2\pi/3$ correspondence. Any L-root can be added to any local c-root pattern to produce a global c-root pattern. Only two combinations are shown.

Each combination generates one type of global c-root pattern which is replicated by rotating it by $2\pi/3$, once CCW and once CW. Global patterns of each type generated by one formula have a π correspondence with those of the same type generated by the other formula. This correspondence makes it possible to easily combine the geometric structures of the two formulas into one fully symmetrical cubic structure.

Combining the two formulas into one generates complications, mainly due to the lack of appropriate algebraic symbolism. Combining the symmetrical structures of the two formulas is more intuitive and much simpler.

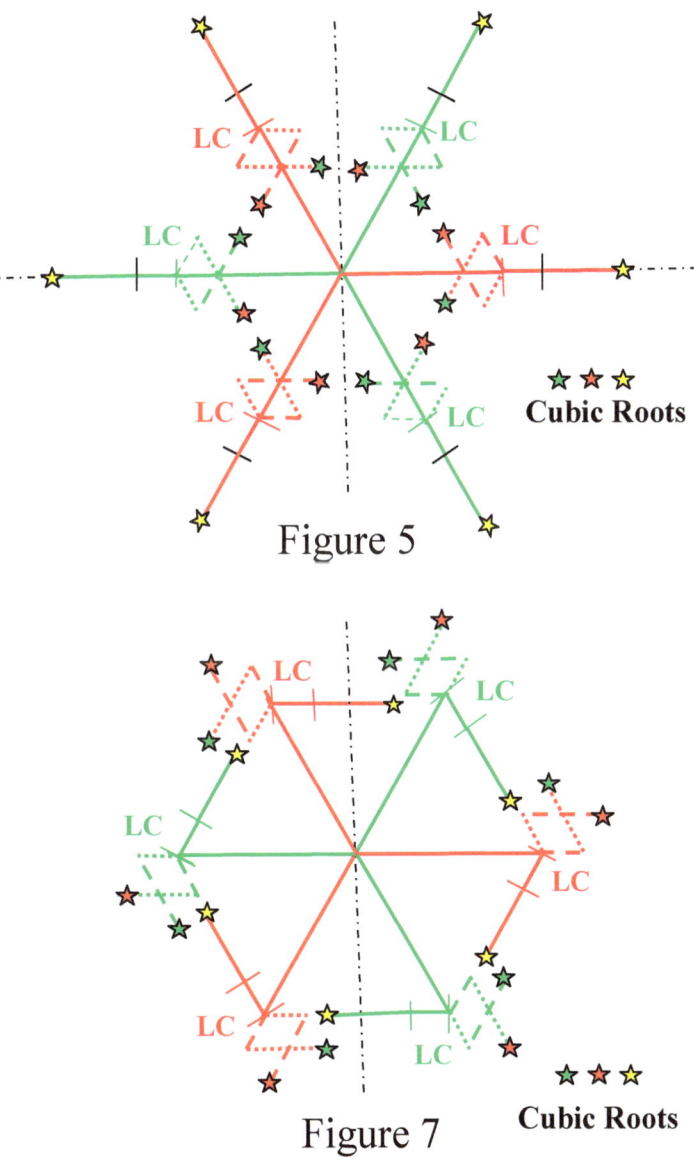

Figure 5

Cubic Roots

Figure 7

Cubic Roots

Choosing to keep the two formulas separate prevents joining the local c-root patterns of one formula to the L-roots of the other and creating fictitious results that are not roots of COI-related cubics. It also limits the number of fully symmetrical cubic structures to three. Each one of these structures contains eighteen cubic roots for a combined total of eighteen times three or fifty-four roots—the same number of roots produced by the two formulas. Hence, all the cubic roots are accounted for.

The COI upgraded and the COI opposite formulas are repeated as a convenience.

$$6y = -2a\sqrt[3]{1} + \sqrt[3]{\left(\frac{C_1^3 + C_2^3}{2}\right) + \sqrt{\left(\frac{C_1^3 - C_2^3}{4}\right)^2}} + \sqrt[3]{\left(\frac{C_1^3 + C_2^3}{2}\right) - \sqrt{\left(\frac{C_1^3 - C_2^3}{4}\right)^2}}$$

$$6y = +2a\sqrt[3]{1} + \sqrt[3]{-\left(\frac{C_1^3 + C_2^3}{2}\right) + \sqrt{\left(\frac{C_1^3 - C_2^3}{4}\right)^2}} + \sqrt[3]{-\left(\frac{C_1^3 + C_2^3}{2}\right) - \sqrt{\left(\frac{C_1^3 - C_2^3}{4}\right)^2}}$$

A scrutiny of the formulas shows that they involve a sequence of operations that need to be performed on essentially the same radical several times, each time with minor changes. This is a common occurrence in software development, and it has simple solutions. Only the first and second radicals of either formula would be used; four sequences of operations would be performed on the chosen radical with flags that uniquely define each sequence. With this approach, there would be just one formula with two radicals, only the needed operations would be executed, and only the proper c-roots would be generated.

Extraneous Solutions

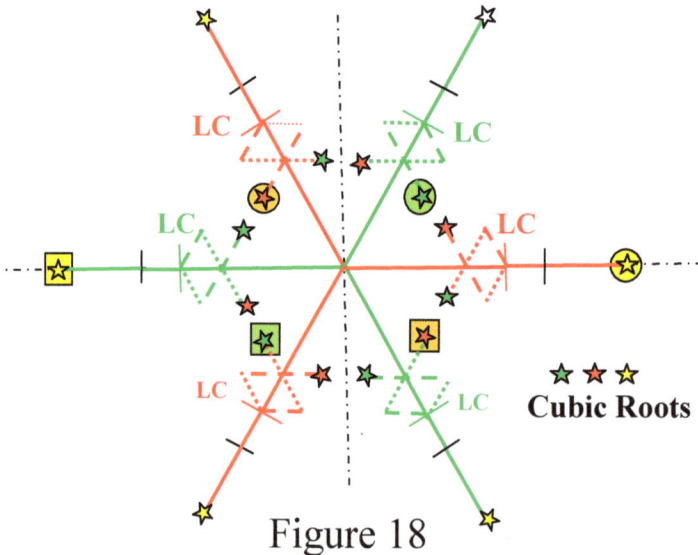

Figure 18

Figure 18 is derived from figure 5. A pair of cubic root triplets has been highlighted to facilitate visualizing the description of additional features.

Consider the symmetrical structure formed by selecting the yellow root from the red, horizontal global pattern, the green root from the green, global pattern at 60°, and the red root from the red, global pattern at 120°. They are highlighted by color-filled circles, as a visual aid.

Consider them as roots of a new cubic. It is evident that the ESF's of this new cubic have values that are neither equal nor symmetrically related to those of the COI. Solving it as an independent cubic will generate the three roots highlighted by the circles, and all other roots related to them by a 60° correspondence, i.e., the very same figure 18.

Roots highlighted by color-filled squares are additive inverses of those highlighted by color-filled circles and are associated with the cubic opposite to the new cubic.

Looking at the cubic formulas (shown on the opposite page), it is seen that every radicand (as a power element) generates a set of three root elements, defined as an R-roots set. A formula root (defined as a Y-root) is obtained by adding together one R-root from each of its radicals, for a total of twenty seven Y-roots per formula. A c-root is obtained by properly selecting three Y-roots. Symmetry among the roots can expedite the process.

All twenty seven combinations allowed by each formula are valid, making the assembly of the total symmetrical structure a rather simple task. But once the structure has been built, it is not so easy to identify a triplet of roots as forming *the* COI roots, especially if the COI equation has complex coefficients. Recall that to make the features of the total cubic symmetric structure easier to be visualized, the parts of figure 18 were purposely chosen to avoid overlapping global patterns, which are to be considered as the norm and not the exception. Also, as discussed earlier, COI real roots can be parts of the family generated by a cubic with complex coefficients.

Roots grouped like those highlighted in figure 18 cannot be roots of any COI related cubic because they mix c-roots derived from both the COI quadratic and the COI opposite quadratic and, as such, they violate the COI opposite quadratics' restriction, which was described on page 19. Root groups of this type will be referred to as *extraneous* roots and their associated equations as *extraneous cubics*. It is evident that the number of extraneous cubics is quite large. Any triplet of non-correspondent c-roots can be chosen as roots of a cubic. There are 6 x 6 x 6 = 216 ways to choose such triplets, with the same triplet chosen six times and the COI triplet being included. Therefore, the structure described by figure 18 is shared by the COI, its five 60° correspondent cubics, and thirty-five extraneous cubics and their five 60° correspondent cubics.

The same is true for the other two cubic families in the COI –related family triplet.

Formulas for the quartic and for the quintic are replete with extraneous roots, but they cannot be readily recognized as such.

Their presence in a quintic formula will be used when discussing Abel's formula for the quintic.

APPENDIX

CONVERTING CUBIC FORMULAS
FROM FUNCTION OF THE ROOTS
TO FUNCTION OF THE COEFFICIENTS

The terms of the various COI formulas were computed as functions of the COI roots and converted to functions of its coefficients via the elementary symmetrical functions (ESFs). The latter are written as functions of the roots and each is related to a specific COI coefficient. The ESFs are designated as σ_1, σ_2, and σ_3 and are as follows:

$$\sigma_1 = x_1 + x_2 + x_0 = -a; \qquad \sigma_2 = x_1x_2 + x_2x_0 + x_0x_1 = b; \qquad \sigma_3 = x_1x_2x_0 = -c$$

The σ_3 ESF relates directly to the COI coefficient c and needs no conversion.

The other terms of the formulas use the symmetrical expressions Σ and Δ^2 as building blocks. These are rewritten as follows:

$$\Sigma = 8[2(x_1^3 + x_2^3 + x_0^3) + 12x_1x_2x_0 - 3(x_1^2x_2 + x_2^2x_0 + x_0^2x_1 + x_1^2x_0 + x_2^2x_1 + x_0^2x_2)] \text{ and}$$

$$\Delta^2 = -1728 \, [(x_1^4x_2^2 + x_2^4x_0^2 + x_0^4x_1^2 + x_1^4x_0^2 + x_2^4x_1^2 + x_0^4x_2^2) +$$

$$+ \, 2x_0x_1x_2(x_1x_2^2 + x_2x_0^2 + x_0x_1^2 + x_1^2x_2 + x_2^2x_0 + x_0^2x_1) +$$

$$-2x_0x_1x_2(x_1^3 + x_2^3 + x_0^3) - 6x_1^2x_2^2x_0^2 - 2(x_1^3x_2^3 + x_2^3x_0^3 + x_0^3x_1^3)]$$

Arithmetic symmetrical expressions are given a rank on the basis of the exponents of the variables present in their terms. A missing variable is assigned an exponent value of zero. The exponents are arranged to maximize their rank values as numbers. The ranks of the three symmetrical expressions for Σ as listed are 300, 111, and 210. The ranks of the five symmetrical expressions for Δ^2 as listed are 420, 321, 411, 222, and 330.

ESFs powers and products usually contain several symmetrical expressions. They are assigned a maximum rank as follows:

$$R(\sigma_1)^K = K00; \qquad R(\sigma_2)^L = LL0; \qquad R(\sigma_3)^M = MMM$$

The symbol $R(\sigma_1)^K$ stands for *the maximum rank of the symmetrical expression σ_1 raised to Kth power*. Similar interpretations are associated with the other symbols.

The conversion procedure involves three steps.

1. The symmetrical expressions of Σ (or Δ^2) are rearranged as a sequence of monotonically decreasing values {MDVs) of their ranks.

2. The power and/or products for ESFs with highest rank equal to the highest ranked expressions in the current form of Σ (or Δ^2) are computed.
3. This power and/or product is then used to replace the current highest-ranked symmetrical expression of Σ (or Δ^2) without affecting the value of Σ (or Δ^2).

As per one above, the ranks of the symmetrical expressions in Σ are rearranged as the MDV sequence 300, 210, and 111 as follows:

$$\Sigma = 8[2(x_1{}^3 + x_2{}^3 + x_0{}^3) - 3(x_1{}^2x_2 + x_2{}^2x_0 + x_0{}^2x_1 + x_1{}^2x_0 + x_2{}^2x_1 + x_0{}^2x_2) + 12x_1x_2x_0]$$

The ESF power with rank 300 (the highest rank currently in Σ) is computed as follows:

$$2\sigma_1{}^3 = 2(x_1 + x_2 + x_0)^3 = 2[(x_1{}^3 + x_2{}^3 + x_0{}^3) + 3(x_1{}^2x_2 + x_2{}^2x_0 + x_0{}^2x_1 + x_1{}^2x_0 + x_2{}^2x_1 + x_0{}^2x_2) + 6(x_1x_2x_0)]$$

Inserting $2\sigma_1{}^3$ into Σ without altering its value—that is by adding and subtracting $2\sigma_1{}^3$, we get:

$$\Sigma = 8[2\sigma_1{}^3 - 9(x_1{}^2x_2 + x_2{}^2x_0 + x_0{}^2x_1 + x_1{}^2x_0 + x_2{}^2x_1 + x_0{}^2x_2)].$$

The highest rank of the expressions now in Σ is now 210. It is computed as follows:

$$-9\sigma_1\sigma_2 = -9(x_1 + x_2 + x_0)(x_1x_2 + x_2x_0 + x_0x_1) =$$
$$= -9[(x_1{}^2x_2 + x_2{}^2x_0 + x_0{}^2x_1 + x_1{}^2x_0 + x_2{}^2x_1 + x_0{}^2x_2) + 3x_1x_2x_0]$$

Adding and subtracting $-9\sigma_1\sigma_2$ to the last value for Σ, we get

$$\Sigma = 8(2\sigma_1{}^3 - 9\sigma_1\sigma_2 + 27x_1x_2x_0) = 8(-2a^3 + 9ab - 27c)$$

The ranks of the symmetrical expressions in Δ^2 are rearranged as the MDV sequence 420, 411, 330, 321, and 222 as follows:

$$\Delta^2 = -1728 [(x_1{}^4x_2{}^2 + x_2{}^4x_0{}^2 + x_0{}^4x_1{}^2 + x_1{}^4x_0{}^2 + x_2{}^4x_1{}^2 + x_0{}^4x_2{}^2) - 2x_0x_1x_2(x_1{}^3 + x_2{}^3 + x_0{}^3) +$$
$$-2(x_1{}^3x_2{}^3 + x_2{}^3x_0{}^3 + x_0{}^3x_1{}^3) + 2x_0x_1x_2(x_1x_2{}^2 + x_2x_0{}^2 + x_0x_1{}^2 + x_1{}^2x_2 + x_2{}^2x_0 + x_0{}^2x_1) +$$
$$-6x_1{}^2x_2{}^2x_0{}^2]$$

The expression for $-9\sigma_1\sigma_2$ has already been found above. From it we get:

$$\sigma_1\sigma_2 = [(x_1{}^2x_2 + x_2{}^2x_0 + x_0{}^2x_1) + (x_1{}^2x_0 + x_2{}^2x_1 + x_0{}^2x_2)] + 3x_1x_2x_0.$$

Then, we compute:

$$(\sigma_1\sigma_2)^2 = (x_1{}^2x_2 + x_2{}^2x_0 + x_0{}^2x_1)^2 + (x_1{}^2x_0 + x_2{}^2x_1 + x_0{}^2x_2)^2 + (3x_1x_2x_0)^2 +$$

$$+ 2(x_1{}^2x_2 + x_2{}^2x_0 + x_0{}^2x_1)(x_1{}^2x_0 + x_2{}^2x_1 + x_0{}^2x_2) +$$
$$+ 6x_1x_2x_0[(x_1{}^2x_2 + x_2{}^2x_0 + x_0{}^2x_1) + (x_1{}^2x_0 + x_2{}^2x_1 + x_0{}^2x_2)]$$

Executing the operations, we get:

$$(\sigma_1\sigma_2)^2 = (x_1{}^4x_2{}^2 + x_2{}^4x_0{}^2 + x_0{}^4x_1{}^2) + 2\,(x_1{}^2x_2{}^3x_0 + x_2{}^2x_0{}^3x_1 + x_0{}^2x_1{}^3x_2) +$$
$$+ (x_1{}^4x_0{}^2 + x_2{}^4x_1{}^2 + x_0{}^4x_2{}^2) + 2(x_1{}^3x_2{}^2x_0 + x_2{}^3x_0{}^2x_1 + x_0{}^3x_1{}^2x_2) + 9x_1{}^2x_2{}^2x_0{}^2 +$$
$$+ 2[(x_1{}^4x_2x_0 + x_2{}^4x_0x_1 + x_0{}^4x_1x_2) + (x_1{}^3x_2{}^3 + x_2{}^3x_0{}^3 + x_0{}^3x_1{}^3) + 3(x_1{}^2x_2{}^2x_0{}^2)] +$$
$$+ 6x_1x_2x_0[x_1{}^2x_2 + x_2{}^2x_0 + x_0{}^2x_1 + x_1{}^2x_0 + x_2{}^2x_1 + x_0{}^2x_2]$$

Rearranging the $(\sigma_1\sigma_2)^2$ terms in MDV sequence, we get:

$$(\sigma_1\sigma_2)^2 = (x_1{}^4x_2{}^2 + x_2{}^4x_0{}^2 + x_0{}^4x_1{}^2 + x_1{}^4x_0{}^2 + x_2{}^4x_1{}^2 + x_0{}^4x_2{}^2) + 2x_1x_2x_0(x_1{}^3 + x_2{}^3 + x_0{}^3) +$$
$$+ 2(x_1{}^3x_2{}^3 + x_2{}^3x_0{}^3 + x_0{}^3x_1{}^3) + 8x_1x_2x_0[x_1{}^2x_2 + x_2{}^2x_0 + x_0{}^2x_1 + x_1{}^2x_0 + x_2{}^2x_1 + x_0{}^2x_2)] +$$
$$+ 15x_1{}^2x_2{}^2x_0{}^2$$

Adding and subtracting $(\sigma_1\sigma_2)^2$ to Δ^2, we get:

$$\Delta^2 = -1728\,[(\sigma_1\sigma_2)^2 - 4x_1x_2x_0(x_1{}^3 + x_2{}^3 + x_0{}^3) - 4(x_1{}^3x_2{}^3 + x_2{}^3x_0{}^3 + x_0{}^3x_1{}^3) +$$
$$- 6x_1x_2x_0\,(x_1{}^2x_2 + x_2{}^2x_0 + x_0{}^2x_1 + x_1{}^2x_0 + x_2{}^2x_1 + x_0{}^2x_2) - 21x_1{}^2x_2{}^2x_0{}^2]$$

To replace the second symmetrical expression (with rank 411), we compute:

$$-4\sigma_1{}^3\sigma_3 = -4[(x_1 + x_2 + x_0)^3x_0x_1x_2] =$$
$$= -4x_0x_1x_2(x_1{}^3 + x_2{}^3 + x_0{}^3) - 12x_1x_2x_0\,(x_1x_2{}^2 + x_2x_0{}^2 + x_0x_1{}^2 + x_1{}^2x_2 + x_2{}^2x_0 + x_0{}^2x_1) +$$
$$- 24x_1{}^2x_2{}^2x_0{}^2$$

Adding and subtracting $-4\sigma_1{}^3\sigma_3$ to the last Δ^2 expression, we get:

$$\Delta^2 = -1728\,[(\sigma_1\sigma_2)^2 - 4\sigma_1{}^3\sigma_3 - 4(x_1{}^3x_2{}^3 + x_2{}^3x_0{}^3 + x_0{}^3x_1{}^3) +$$
$$+ 6x_1x_2x_0\,(x_1{}^2x_2 + x_2{}^2x_0 + x_0{}^2x_1 + x_1{}^2x_0 + x_2{}^2x_1 + x_0{}^2x_2) + 3x_1{}^2x_2{}^2x_0{}^2]$$

To replace the third symmetrical expression (with rank 330), we compute:

$$-4\sigma_2{}^3 = -4(x_1x_2 + x_2x_0 + x_0x_1)^3 = -4(x_1{}^3x_2{}^3 + x_2{}^3x_0{}^3 + x_0{}^3x_1{}^3) +$$
$$- 12(x_1{}^2x_2{}^3x_0 + x_2{}^2x_0{}^3x_1 + x_0{}^2x_1{}^3x_2 + x_1{}^3x_2{}^2x_0 + x_2{}^3x_0{}^2x_1 + x_0{}^3x_1{}^2x_2) - 24x_1{}^2x_2{}^2x_0{}^2$$

Adding and subtracting $\sigma_2{}^3$ to the last expression for Δ^2, we get:

$$\Delta^2 = -1728\,[(\sigma_1\sigma_2)^2 - 4\sigma_1{}^3\sigma_3 - 4\sigma_2{}^3 + 18x_0x_1x_2\,(x_1x_2{}^2 + x_2x_0{}^2 + x_0x_1{}^2 + x_1{}^2x_2 + x_2{}^2x_0 + x_0{}^2x_1) +$$
$$+ 27x_1{}^2x_2{}^2x_0{}^2]$$

The expression for $\sigma_1\sigma_2$ was found to be the following:

$$\sigma_1\sigma_2 = (x_1^2x_2 + x_2^2x_0 + x_0^2x_1 + x_1^2x_0 + x_2^2x_1 + x_0^2x_2) + 3x_1x_2x_0$$

From it, we get:

$$18\,\sigma_1\sigma_2\sigma_3 = 18(x_1^2x_2 + x_2^2x_0 + x_0^2x_1 + x_1^2x_0 + x_2^2x_1 + x_0^2x_2 + 3x_1x_2x_0)\, x_1x_2x_0 =$$
$$= 18x_1x_2x_0(x_1^2x_2 + x_2^2x_0 + x_0^2x_1 + x_1^2x_0 + x_2^2x_1 + x_0^2x_2) + 54x_1^2x_2^2x_0^2$$

The latest expression for Δ^2 is repeated for convenience

$$\Delta^2 = -1728\,[(\sigma_1\sigma_2)^2 - 4\sigma_1^3\sigma_3 - 4\sigma_2^3 + 18x_0x_1x_2\,(x_1x_2^2 + x_2x_0^2 + x_0x_1^2 + x_1^2x_2 + x_2^2x_0 + x_0^2x_1) +$$
$$+27x_1^2x_2^2x_0^2]$$

Adding and subtracting $18\,\sigma_1\sigma_2\sigma_3$ into it finally we get:

$$\Delta^2 = -1728\,[(\sigma_1\sigma_2)^2 - 4\sigma_1^3\sigma_3 - 4\sigma_2^3 + 18\,\sigma_1\sigma_2\sigma_3 - 27\sigma_3^2] =$$
$$= -1728\,[a^2b^2 - 4a^3c - 4b^3 + 18abc - 27c^2]$$

Computations needed to change the COI formula from functions of symmetrical expressions of the roots to function of its coefficient have been made according to normally used procedures. They are hard to follow and prone to errors, especially those for Δ^2. There is a better and simpler way. For many steps of the actual procedure several symmetrical expressions remain invariant and can be replaced by a constant such as their rank from which they are readily and easily reestablished to their expanded form. Additionally, it is possible to give the computations an easy-to-visualize tabular form.

The expressions for Δ^2, $(\sigma_1\sigma_2)^2$, $-4\sigma_1^3\sigma_3$, and $-4\sigma_2^3$ are repeated for convenience.

$$\Delta^2 = -1728\,[(x_1^4x_2^2 + x_2^4x_0^2 + x_0^4x_1^2 + x_1^4x_0^2 + x_2^4x_1^2 + x_0^4x_2^2) - 2x_0x_1x_2(x_1^3 + x_2^3 + x_0^3) +$$
$$-2(x_1^3x_2^3 + x_2^3x_0^3 + x_0^3x_1^3) + 2x_0x_1x_2(x_1x_2^2 + x_2x_0^2 + x_0x_1^2 + x_1^2x_2 + x_2^2x_0 + x_0^2x_1) +$$
$$-6x_1^2x_2^2x_0^2]$$

$$(\sigma_1\sigma_2)^2 = (x_1^4x_2^2 + x_2^4x_0^2 + x_0^4x_1^2 + x_1^4x_0^2 + x_2^4x_1^2 + x_0^4x_2^2) + 2x_1x_2x_0(x_1^3 + x_2^3 + x_0^3) +$$
$$+2(x_1^3x_2^3 + x_2^3x_0^3 + x_0^3x_1^3) + 8x_1x_2x_0[x_1^2x_2 + x_2^2x_0 + x_0^2x_1 + x_1^2x_0 + x_2^2x_1 + x_0^2x_2)] +$$
$$+15x_1^2x_2^2x_0^2$$

$$-4\sigma_1^3\sigma_3 = -4x_0x_1x_2(x_1^3 + x_2^3 + x_0^3) - 12x_1x_2x_0\,(x_1x_2^2 + x_2x_0^2 + x_0x_1^2 + x_1^2x_2 + x_2^2x_0 + x_0^2x_1) +$$
$$-24x_1^2x_2^2x_0^2$$

$$-4\sigma_2^3 = -4\,(x_1^3x_2^3 + x_2^3x_0^3 + x_0^3x_1^3) +$$
$$-12x_0x_1x_2[(x_1x_2^2 + x_1^2x_2) + (x_2x_0^2 + x_2^2x_0) + (x_0x_1^2 + x_1x_0^2)] - 24x_1^2x_2^2x_0^2$$

Let $\Delta_1^2 = \dfrac{\Delta^2}{-1728}$

The operations needed to convert the COI formula to a function of the coefficients can be given the following tabular form.

	420	411	330	321	222
Δ_1^2	1	-2	-2	+2	-6
$(\sigma_1\sigma_2)^2$	1	+2	+2	+8	+15
Residue 1	$(\sigma_1\sigma_2)^2$	-4	-4	-6	-21
$-4\sigma_1{}^3\sigma_3$	0	-4	0	-12	-24
Residue 2	$(\sigma_1\sigma_2)^2$	$-4\sigma_1{}^3\sigma_3$	-4	+6	+3
$-4\sigma_2{}^3$	0	0	-4	-12	-24
Residue 3	$(\sigma_1\sigma_2)^2$	$-4\sigma_1{}^3\sigma_3$	$-4\sigma_2{}^3$	+18	+27
$18\sigma_1\sigma_2\sigma_3$	0	0	0	+18	+54
Residue 4	$(\sigma_1\sigma_2)^2$	$-4\sigma_1{}^3\sigma_3$	$-4\sigma_2{}^3$	$+18\sigma_1\sigma_2\sigma_3$	$-27\sigma_3{}^2$

In the table above, the top row shows the ranks of symmetrical expressions in MDV sequence. The first column lists the quantities involved also in MDV sequence. The rest of the table shows the coefficient associated with a specific rank of the quantity heading the row. For instance, the second and third rows display the coefficients of Δ_1^2 and of $(\sigma_1\sigma_2)^2$ respectively.

The fourth row (Residue 1) shows the result of adding and subtracting $(\sigma_1\sigma_2)^2$ to Δ_1^2. For every column with rank lower than 420, the sum of the coefficients in the rows headed by Residue 1 and $(\sigma_1\sigma_2)^2$ is equal to the coefficient of Δ_1^2.

The process continues by adding and subtracting $-4\sigma_1{}^3\sigma_3$ to the row headed by Residue 1 to obtain the Residue 2 row. For every column with rank lower than 411, the sum of the coefficients in the rows headed by Residue 2 and $-4\sigma_1{}^3\sigma_3$ is equal to the coefficient of the Residue 1 row.

Similar procedures are used to obtain the Residue 3 row by adding and subtracting $-4\sigma_2{}^3$ to the Residue 2 row to obtain the Residue 3 row and then adding and subtracting $18\sigma_1\sigma_2\sigma_3$ (shown on the opposite page) into the Residue 3 row to obtain the Residue 4 row.

It is interesting to note that adding a zero sum quantity is a frequently occurring procedure in dealing with symmetrical structures and that fully symmetrical structures are themselves zero sum entities.

ABEL'S PROOF

After completing most of his work, the author sought to enlist the help of other professionals but found that they were very skeptical about his claim that he had found a method for solving the quintic. Several mathematicians told him that he had probably made a mistake somewhere along the way in his development and that the task had been proven to be impossible. A family friend suggested that he should read and understand Abel's proof and suggested the following three books as references:

Abstract Algebra by I. N. Herstein

Abstract Algebra and Solutions by Radicals by John E. Maxfield and Margaret W. Maxfield

Abel's Proof: An Essay on the Meaning of Mathematical Unsolvability by Peter Pesic

The author added two more to the list:

The Equation That Couldn't Be Solved, How Mathematical Genius Discovered the Language of Symmetry by Mario Livio

Supersymmetry: Unveiling The Ultimate Laws Of Nature by Gordon Kane

The first book just mentions Abel's name. The second and fourth books are about Abel's proof, but they do not provide any details on the proof itself. The fifth book was one of those used to search for an academic definition of symmetry.

The third book provides annotated translations of Abel's original 1824 and 1826 papers as appendix A and appendix B, respectively. I thoroughly enjoyed reading the book for the insights that it provides on many mathematical concepts.

Only the relevant points of Pesic's rendition of Abel's original 1824 paper are presented here for the sake of brevity.

Abel uses fractional exponents to indicate radicals. Here, they are written using the more conventional, contemporary notation. For instance,

$$\sqrt[5]{R} \rightarrow R^{\frac{1}{5}}$$

The symbol $\rightarrow$ indicates that the first term replaces the second.

Additionally, as Pesic remarks, Abel's notation for the coefficients does not follow the conventions more commonly used today.

Abel's assumptions are as follows:

Let

$$y^5 - ay^4 + by^3 - cy^2 + dy - e = 0$$

be the general equation of the fifth degree and let us suppose that it is solvable algebraically, that is, one can express y by a function formed by radicals of the quantities a, b, c, d, and e.

It is clear that, in this case, we can express y in the form

$$y = p + p_1 \sqrt[m]{R} + p_2 \sqrt[m]{R^2} + \cdots + p_{m-1} \sqrt[m]{R^{m-1}}$$

m being a prime number and R, p, p_1, p_2, etc. functions of the same form as y and so on, until we come to rational functions of the quantities a, b, c, d, and e.

Note: there seems to be a discrepancy here. The generic prime number m has replaced the prime number five. This notation appears in Abel's 1826 paper in which he extended his proof to any equation with a generic degree m, with m being a prime number. It seems safe here to replace m with five and focus on the quintic, which is rewritten as follows:

$$y = p + p_1 \sqrt[5]{R} + p_2 \sqrt[5]{R^2} + p_3 \sqrt[5]{R^3} + p_4 \sqrt[5]{R^4}$$

The quintic above will be referred to as the *quintic of interest* (QOI).

To avoid ambiguities, the formula's solutions will be referred to as *Y-roots*, the roots of the four radicals as *R-roots*, and the five roots of the QOI as *Q-roots*. The latter are a very small subset of *Y-roots*.

In his proof, Abel uses variable substitutions that include inserting Q-roots back into the original equation. He also uses Group Theory theorems in his proof. The reasons why expressions containing radicals fall outside the Group Theory domain were discussed earlier in this book.

A remark is apropos here. Creating a formula involves raising some quantities to a power. The result of this operation is unique, and that makes it look like a legitimate Group Theory operation. But it comes with a hidden side effect: the correspondent parts of said quantities' are surreptitiously inserted into the result. Computing Y-roots involves extracting roots from these previously generated powers: the result of this operation contains all the previously inserted correspondent parts, any one of which *can* and one *has* to be chosen. In general, there will be no clues as to which Y-root will produce a desired objective.

All radicals in Abel's formula produce a minimum of five roots, more if the radicand R contains nested radicals, which are allowed under his assumptions. To simplify things, assume that p_1, p_2, p_3, and p_4 have a single value so that only the second, third, fourth, and fifth term of the formula generate a multiplicity of terms.

Together, these four terms produce six hundred and twenty five combinations for *the R*-roots (five raised to the fourth power). Each combination when added to the quantity *p* generates a *Y*-root. Five of them are Q-roots. The probability for choosing a first Q-root is one in 625. The probability for choosing the other four Q-roots is one in 624 for the second, one in 623 for the third, one in 622 for the fourth, and one in 621 for the fifth. The probability for choosing all five Q-roots is then one in 625 times 624 times 623 times 622 times 621. By comparison, 500 (which is less than all five numbers above) raised to the fifth power is equal to 3125 followed by ten zeroes,—a number with fourteen decimal places.

In general, the formula does not provide any clues on how to choose a *Y*-root, which is also a *Q*-root.

One way to identify all five Q-roots is to add together a combination of five *Y*-roots and then to verify if their sum is equal to the coefficient *a* (using Abel's own notation). This procedure is analogous to what was done with the original cubic formula but it is much bigger, more complex and laborious.

If the quantity *p* also has five values, the number of Y-root combinations increases to three thousands one hundred and twenty five (five raised to the fifth power), while the number of combinations that produce Q-roots stays the same.

If *R* contained just one *nested radical* with index four, then the number of roots per radical would be twenty: five radical roots for each root of its nested radical. Together, the formula last four terms would produce 160,000 combinations of valid Y-roots (twenty raised to the fourth power) five of which are Q-roots. Another way to verify that a Y-root is also a Q-root is to substitute it for *y* in the equation and see if the result is equal to zero.

If *R* also contained radicals with index three and two, then the number of combinations would be correspondingly much larger.

If the quantity *p* also had five values and *R* contained one nested radical with index four, then the total number of combinations that produce a valid Y-root would increase to eight hundred thousand (160,000 times five). Finding the five Q-roots would still require adding a group of five *Y*-roots and verifying that their sum is the additive inverse of the coefficient *a*, or substituting a Y-root for the variable y into the equation and verifying that the result is zero.

Abel's proof seems to make the unstated assumption that the presumed quintic formula generates only five values for the *Y*-roots, all of which would be also Q-roots.

Incidentally, Abel did find a solution for the quintic. His credo (Pesic's book page 151) was, "One must give to a problem a form such that it is always possible to solve it, which one can always do with any problem. Instead of looking for a relation that one does not know exists or not, one must ask if such relation is really possible."

To solve the quintic, Abel used elliptic functions.

On page 91 of his book, Pesic remarks that the quadratic formula does not have Abel's form. He makes some substitutions and succeeds in giving the formula a form that is consistent with Abel's general form. In a sense, he confirms the fact that an equation and its formula can be put in many different forms.

Previous methods for finding formulas work with the equation coefficients, change their form, and make assumptions searching for one that works. With the symmetrical approach, DyConN matrices (with N an appropriate positive integer) keep symmetrically related groups of roots together as local structures. Then, as progress is made, deficiencies (mainly in the form of parts without correspondent ones) are easily spotted and corrected. No unproven assumptions need to be made at any time during the development.

A way for analyzing the features of a generic quintic and for getting a glimpse of its magnitude and complexity is to use its ESFs. They are designated as σ_{51}, σ_{52}, σ_{53}, σ_{54}, and σ_{55}.

The generic QOI is rewritten below for convenience. It uses the common modern notation.

$$y^5 + ay^4 + by^3 + cy^2 + dy + e = 0$$

Let y_1, y_2, y_3, y_4, and y_5 be its roots. Then the QOI can be given the following form:

$$(y - y_1)(y - y_2)(y - y_3)(y - y_4)(y - y_5) = 0$$

Performing the above operations, we get

$$
\begin{aligned}
y^5 &- (y_1 + y_2 + y_3 + y_4 + y_5)y^4 \\
&+ (y_1y_2 + y_1y_3 + y_1y_4 + y_1y_5 + y_2y_3 + y_2y_4 + y_2y_5 + y_3y_4 + y_3y_5 \\
&+ y_4y_5)y^3 \\
&- (y_1y_2y_3 + y_1y_2y_4 + y_1y_2y_5 + y_1y_3y_4 + y_1y_3y_5 + y_1y_4y_5 + y_2y_3y_4 \\
&+ y_2y_3y_5 + y_2y_4y_5 + y_3y_4y_5)y^2 \\
&+ (y_1y_2y_3y_4 + y_1y_2y_3y_5 + y_1y_2y_4y_5 + y_1y_3y_4y_5 + y_2y_3y_4y_5)y \\
&- y_1y_2y_3y_4y_5 = 0
\end{aligned}
$$

Comparing both quintic forms above, we get the relations between ESF's and the quintic coefficients

$$\sigma_{51} = y_1 + y_2 + y_3 + y_4 + y_5 = -a$$

$$\sigma_{52} = y_1y_2 + y_1y_3 + y_1y_4 + y_1y_5 + y_2y_3 + y_2y_4 + y_2y_5 + y_3y_4 + y_3y_5 + y_4y_5 = b$$

$$
\begin{aligned}
\sigma_{53} = y_1y_2y_3 &+ y_1y_2y_4 + y_1y_2y_5 + y_1y_3y_4 + y_1y_3y_5 + y_1y_4y_5 + y_2y_3y_4 + y_2y_3y_5 \\
&+ y_2y_4y_5 + y_3y_4y_5 = -c
\end{aligned}
$$

$$\sigma_{54} = y_1y_2y_3y_4 + y_1y_2y_3y_5 + y_1y_2y_4y_5 + y_1y_3y_4y_5 + y_2y_3y_4y_5 = d$$

$$\sigma_{55} = y_1 y_2 y_3 y_4 y_5 = -e$$

The value of σ_{51} is unique to the QOI. As such, it cannot be part of any symmetrical structure and/or symmetrical formula, because it has no correspondent parts.

Terms consisting of a plurality of Y-roots are defined as corresponding terms if they have the same value and contain Y-roots that have the same subscripts. Examples are

$y_1 y_2$ and $(-y_1)(-y_2)$; $y_1 y_2 y_3$ and $y_1(\omega_1 y_2)(\omega_2 y_3)$ and $(\omega_2 y_1) y_2 (\omega_1 y_3)$

The value of σ_{52} is not unique to the QOI, but it is shared with another quintic designated as the *opposite quintic*. Their roots are bound by a π correspondence. Their b coefficients are equal, as shown below.

$$\begin{aligned}
\sigma_{52} &= y_1 y_2 + y_1 y_3 + y_1 y_4 + y_1 y_5 + y_2 y_3 + y_2 y_4 + y_2 y_5 + y_3 y_4 + y_3 y_5 + y_4 y_5 \\
&= (-y_1)(-y_2) + (-y_1)(-y_3) + (-y_1)(-y_4) + (-y_1)(-y_5) + (-y_2)(-y_3) \\
&\quad + (-y_2)(-y_4) + (-y_2)(-y_5) + (-y_3)(-y_4) + (-y_3)(-y_5) + (-y_4)(-y_5) \\
&= b
\end{aligned}$$

Together, they are defined as a *quintic pair*. Their roots are parts of a symmetrical structure and are bound by a π correspondence. A symmetrical formula would generate Y-roots for both quintics in the pair. Note that because of the π correspondence the quintics in the pair cannot share any Y-roots.

There seems to be a paradox here. A symmetrical formula should generate Q-roots for both the QOI and its opposite quintic, but it must also keep them separate. The situation is similar to that discussed for the COI and its opposite cubic (page 52).

Every term of σ_{53} is valid not only for the Q-root triplets, as shown in the opposite page, but also for the root triplets of two additional quintics, as shown by the following ten equalities:

$$y_1 y_2 y_3 = (\omega_1 y_1)(\omega_1 y_2)(\omega_1 y_3) = (\omega_2 y_1)(\omega_2 y_2)(\omega_2 y_3)$$

$$y_1 y_2 y_4 = (\omega_1 y_1)(\omega_1 y_2)(\omega_1 y_4) = (\omega_2 y_1)(\omega_2 y_2)(\omega_2 y_4)$$

$$y_1 y_2 y_5 = (\omega_1 y_1)(\omega_1 y_2)(\omega_1 y_5) = (\omega_2 y_1)(\omega_2 y_2)(\omega_2 y_5)$$

$$y_1 y_3 y_4 = (\omega_1 y_1)(\omega_1 y_3)(\omega_1 y_4) = (\omega_2 y_1)(\omega_2 y_3)(\omega_2 y_4)$$

$$y_1 y_3 y_5 = (\omega_1 y_1)(\omega_1 y_3)(\omega_1 y_5) = (\omega_2 y_1)(\omega_2 y_3)(\omega_2 y_5)$$

$$y_1 y_4 y_5 = (\omega_1 y_1)(\omega_1 y_4)(\omega_1 y_5) = (\omega_2 y_1)(\omega_2 y_4)(\omega_2 y_5)$$

$$y_2 y_3 y_4 = (\omega_1 y_2)(\omega_1 y_3)(\omega_1 y_4) = (\omega_2 y_2)(\omega_2 y_3)(\omega_2 y_4)$$

$$y_2 y_3 y_5 = (\omega_1 y_2)(\omega_1 y_3)(\omega_1 y_5) = (\omega_2 y_2)(\omega_2 y_3)(\omega_2 y_5)$$

$$y_2 y_4 y_5 = (\omega_1 y_2)(\omega_1 y_4)(\omega_1 y_5) = (\omega_2 y_2)(\omega_2 y_4)(\omega_2 y_5)$$

$$y_3y_4y_5 = (\omega_1 y_3)(\omega_1 y_4)(\omega_1 y_5) = (\omega_2 y_3)(\omega_2 y_4)(\omega_2 y_5)$$

The sum of root triplets in every column is equal to σ_{53}. Those in the second and third columns are associated with two quintics that, together with the QOI, will be referred to as the *QOI quintic triplet*. Their corresponding Y-root triplets are bound by a $2\pi/3$ correspondence, and as such, they are parts of a symmetrical structure. Again, note that because of this correspondence the quintics in the triplet cannot share any Y-roots.

If the Q-roots are replaced by corresponding ones from the opposite quintic, they generate another quintic triplet (referred to as the *opposite quintic triplet*) with root triplets that have a $2\pi/3$ correspondence within the opposite quintic triplet and a π correspondence with those of the QOI quintic triplet. As a group, the six quintics have root triplets bound by a $\pi/3$ correspondence, and they form a symmetric structure. A symmetrical formula would have to generate the roots of all six quintics.

Q-root quadruplets in σ_{54} are bound by a $\pi/2$ correspondence to those of three more quintics, as shown below.

$$y_1y_2y_3y_4 = (jy_1)(jy_2)(jy_3)(jy_4) = (j^2 y_1)(j^2 y_2)(j^2 y_3)(j^2 y_4) = (j^3 y_1)(j^3 y_2)(j^3 y_3)(j^3 y_4)$$

$$y_1y_2y_3y_5 = (jy_1)(jy_2)(jy_3)(jy_5) = (j^2 y_1)(j^2 y_2)(j^2 y_3)(j^2 y_5) = (j^3 y_1)(j^3 y_2)(j^3 y_3)(j^3 y_5)$$

$$y_1y_2y_4y_5 = (jy_1)(jy_2)(jy_4)(jy_5) = (j^2 y_1)(j^2 y_2)(j^2 y_4)(j^2 y_5) = (j^3 y_1)(j^3 y_2)(j^3 y_4)(j^3 y_5)$$

$$y_1y_3y_4y_5 = (jy_1)(jy_3)(jy_4)(jy_5) = (j^2 y_1)(j^2 y_3)(j^2 y_4)(j^2 y_5) = (j^3 y_1)(j^3 y_3)(j^3 y_4)(j^3 y_5)$$

$$y_2y_3y_4y_5 = (jy_2)(jy_3)(jy_4)(jy_5) = (j^2 y_2)(j^2 y_3)(j^2 y_4)(j^2 y_5) = (j^3 y_2)(j^3 y_3)(j^3 y_4)(j^3 y_5)$$

The sum of the terms of every column above is equal to the QOI d coefficient. These four quintics will be referred to as the *QOI quintic quadruplet*. Their roots are parts of a symmetrical structure, and any eventual quintic symmetrical formula will produce all of them.

Quadruplets of the opposite quintic are obtained by replacing every Q-root with its additive inverse. For instance

$$(jy_1)(jy_2)(jy_3)(jy_4) \Leftarrow [(j)(-y_1)][(j)(-y_2)][(j)(-y_3)][(j)(-y_4)]$$

Where the symbol $\Leftarrow$ is shorthand for "the second term replaces the first."

The replacement generates four new quintics denominated as the *QOI opposite quintic quadruplet*. Root quadruplets within this quadruplet have a $\pi/2$ correspondence among themselves and a π correspondence with those of the QOI quintic quadruplet.

Note that it is possible for these quadruplets to share some Y-roots although structures with roots bound by any symmetry cannot share any. This abnormality between the roots of the QOI quintic quadruplet and those of the QOI opposite quintic quadruplet is due to the fact that four is not a prime number. Clues about how to handle it are provided by the quartic structure.

Define as $\{\rho_1, \rho_2, \rho_3, \rho_4, \rho_0\}$ the five principal fifth roots of unity. Then σ_{55} can be expressed as follows:

$$\sigma_{55} = y_1 y_2 y_3 y_4 y_5 = (\rho_1 y_1)(\rho_1 y_2)(\rho_1 y_3)(\rho_1 y_4)(\rho_1 y_5) = (\rho_2 y_1)(\rho_2 y_2)(\rho_2 y_3)(\rho_2 y_4)(\rho_2 y_5)$$
$$= (\rho_3 y_1)(\rho_3 y_2)(\rho_3 y_3)(\rho_3 y_4)(\rho_3 y_5) = (\rho_4 y_1)(\rho_4 y_2)(\rho_4 y_3)(\rho_4 y_4)(\rho_4 y_5)$$

The preceding equalities show that the QOI e coefficient is shared by at least five quintics, the roots of which are bound by a $2\pi/5$ correspondence. Together, they are designated as the *QOI quintuplet*.

Quintuplets of the opposite quintic are obtained by replacing every Q-root with its additive inverse. These quintics are denominated as the *QOI opposite quintic quintuplet*. Quintics in the QOI quintuplet have a $2\pi/5$ among themselves and a π correspondence with the roots of the QOI opposite quintics. Together, they are parts of a symmetrical structure.

The conclusion is that a fully symmetrical quintic formula and a fully symmetric quintic structure are extremely large and very complex.

At this point, whether all the Y-roots are parts of one or more symmetrical quintic formulas or some are to be considered as extraneous solutions (see figure 18, page 52) can only be a matter of speculation.

The $2\pi/5$ correspondence of the QOI quintuplet and its relationships to the π, $2\pi/3$ and $\pi/2$ correspondences define the quintic structure's overall correspondence. The difference between $\pi/2$ and $2\pi/5$ is equal to $\pi/10$. It is the smallest difference among all correspondence angles, and it might be the overall correspondence among the major parts of its symmetrical structure.

The author has done enough work to show that the fully symmetrical quintic structure can be assembled if sufficient resources are dedicated to it.

So far, the only symmetrical structures considered have been the basic ones—that is, those with parts that are roots of some quintic symmetrically related to the QOI. As it happened with the total cubic symmetric structure, parts of the fully symmetrical quintic structure are not limited to those considered. Many of them form local symmetric structures that need to be joined with others to form larger local and/or global symmetric structures. The parts of local symmetric structures of the same type have the same degree, which is an appropriate power of five. At this time, identifying them can only be matter of speculation.

The author has expanded to the quintic the methodology used to find formulas and to assemble fully symmetrical structures for the cubic and the quartic.

Work has been completed to develop a procedure that creates formulas for the quintic and would assemble them as a fully symmetrical structure, provided that adequate resources are dedicated to the task.

The results of this work will be the subject of future books.

EPILOGUE

NEXT PROJECT

A methodology similar to that used for the cubic but with many additional features has been applied to search for the formulas of both the quartic and quintic equations. The entire fully symmetrical formulas have been successfully generated for the quartic as functions of both its roots and of its coefficients. Work for assembling the formulas as geometric symmetrical structures is yet to be completed.

Enough work has been completed to develop a detailed procedure that will create a fully symmetrical formulas and structures for the quintic provided that sufficient resources are devoted to the problem. Needed resources far exceed what any individual contributor can muster.

The quartic symmetrical structure contains a large number of fully symmetrical cubic structures that inherit all the features of the fully symmetrical cubic structures. The quartic structure has additional features that are not found in the symmetrical cubic structure. These additional features and their implications are outlined next.

The first feature derives from the fact that four is not a prime number. Because of this, the problem of finding the quartic formulas breaks down into several separate, different, but interdependent paths, which complicate the correspondence relationships that need to exist among all the similar parts of any symmetrical structure.

A second problem is the sheer size and complexity of the total quartic symmetrical structure. As the development of the formulas progresses, the size and the number of parts involved (roots and symmetrical combinations thereof) range from one decimal place to twenty-two, whereas in comparison, those related to the cubic structure range from one to three decimal places. To obtain accurate results and to present them in a reasonably compact and user-friendly format, it was necessary to develop a special symbolism to compress their representation to the absolutely indispensable elements. The translation of Σ and Δ^2 from functions of the cubic roots to functions of the cubic coefficients is an example of how laborious, prone to error, and overloaded with unnecessary symbols the commonly used processes are. Many of the symbols in any symmetric expression are essentially invariant placeholders that do not need to be carried along every step of the way—symmetry makes this feature possible.

An area that needed improvements over the common practice was computing the products and sums of expressions containing very large numbers of super and/or subscripted variables. A compact, bare-bones notation became a necessity and a great asset in creating reusable and expandable software routines that fully automate all computations. It also contributed to optimizing the presentation of the results and to make it as user friendly as possible.

Current plans are to write a second book covering the work done on the quartic and the quintic.

The quartic formulas have been fully developed, but their documentation is very poor and fragmented. It will require much work before it can be assembled into a coherent, user-friendly compendium of information. A book format is not the best medium for conveying such information. A CD is a better choice. With it and with the help of a computer, large amounts of information can be displayed simultaneously, over several monitors, for a better and more comprehensive evaluation of how the various parts play together to form symmetrical structures.

The author will need much help to turn this project into reality.

ABOUT THE AUTHOR

Carlo Faustini was born in Italy, where he received his initial education from the elementary to undergraduate levels. His preferred subjects were mathematics and Latin. He received a BSEE from Rutgers University and an MSEE degree from the University of Pennsylvania. He worked for several companies, mostly doing classified work related to national defense. He also took many continuing education courses to keep up with the state of the art.

His ambitions were to do applied research, but he realized that his qualifications and the type of work he was involved with offered very scant opportunities. During what seemed a never-ending period of very dull work, he decided to seek a research project of his own, one that could involve a modicum of original research and that he could do within his meager resources.

In his last years of high school, he became very familiar with the development, characteristics, and uses of the quadratic formula and learned that one also existed for the cubic. During his undergraduate years, he came across problems that required finding the roots of high-degree equations for which no formulas existed. So he set out to find a formula for the quartic as the goal of his research project.

To keep the research original, Faustini avoided previous work on the subject. To broaden the project's scope, he expanded the goal to searching for a generic methodology that could be applied to equations of any degree. He spent more than a decade and considerable effort in searching for the essential characteristics that define the problem to ensure that his solutions addressed the real problem and not preferential solutions. It took a lot of time and effort to choose the degree for the first equation to be solved. He finally decided on the cubic because he had some vague ideas regarding how to go about it and knew that, if the process was successful, he could compare the results of his efforts with the existing formula, with which he was not familiar at the time.

He found a first cubic formula that is essentially the same as Cardano's 1545 formula. At this point, he looked for partners and found that a formula for the quartic had already been developed, and as for the quintic, there were proofs that it could not be solved. Most relevant among them is one developed by the famous mathematician Abel.

Cardano's and Faustini's first formula have two faults. They are not symmetrical because one of their terms does not have correspondent ones and both produce nine solutions, of which only three can be associated with the roots of the cubic of interest.

Moving on to the quartic, he found that the strategy used for the cubic would not succeed. Suspecting that the reason was due to four not being a prime number, he started work on the quintic, searching for clues. He found that the quintic shares the same problem as the quartic, plus it has many new ones of its own. Discovering the common problem helped solve the quartic.

Mr. Faustini's initial literature searches were limited to the bare essentials. After most of his work had been done, he perused the books listed in the section about Abel's proof to reconcile his results with what had been done previously.